Science: Philosophy, History and Education

Scope of the Series

This book series serves as a venue for the exchange of the complementary perspectives of science educators and HPS scholars. History and philosophy of science (HPS) contributes a lot to science education and there is currently an increased interest for exploring this relationship further. Science educators have started delving into the details of HPS scholarship, often in collaboration with HPS scholars. In addition, and perhaps most importantly, HPS scholars have come to realize that they have a lot to contribute to science education, predominantly in two domains: a) understanding concepts and b) understanding the nature of science. In order to teach about central science concepts such as “force”, “adaptation”, “electron” etc, the contribution of HPS scholars is fundamental in answering questions such as: a) When was the concept created or coined? What was its initial meaning and how different is it today? Accordingly, in order to teach about the nature of science the contribution of HPS scholar is crucial in clarifying the characteristics of scientific knowledge and in presenting exemplar cases from the history of science that provide an authentic image of how science has been done. The series aims to publish authoritative and comprehensive books and to establish that HPS-informed science education should be the norm and not some special case. This series complements the journal Science & Education http://www.springer.com/journal/11191Book Proposals should be sent to the Publishing Editor at bernadette.ohmer@springer.com.

More information about this series at http://www.springer.com/series/13387

Mansoor Niaz

Chemistry Education and Contributions from History and Philosophy of Science

Springer

Mansoor Niaz
Epistemology of Science Group
Department of Chemistry
Universidad de Oriente
Cumaná, Estado Sucre, Venezuela

Science: Philosophy, History and Education
ISBN 978-3-319-26246-8 ISBN 978-3-319-26248-2 (eBook)
DOI 10.1007/978-3-319-26248-2

Library of Congress Control Number: 2015958770

Springer Cham Heidelberg New York Dordrecht London

Printed on acid-free paper

Springer International Publishing AG Switzerland is part of Springer Science+Business Media (www.springer.com)

For Magda and Sabuhi

Foreword

The idea of relying on the history and philosophy of science in order to facilitate the teaching and learning of science is not new. Nevertheless, as this book by Mansoor Niaz shows, we are still far from making a good use of what history and philosophy of science can contribute to chemistry education (I can confirm that this is the case for biology education, too). As several scholars have repeatedly noted, history and philosophy of science provide solid foundations and frameworks for developing teaching materials that would facilitate understanding of science concepts and nature of science. However, teachers have difficulties in mastering this kind of material, and textbook authors usually do not include much historical information besides biographical details and historical vignettes. As a result the teaching of science is quite often ahistorical, in the sense that it does not provide students with an authentic view of how scientific concepts were created, how empirical evidence was acquired in order to support or reject hypotheses, or how scientific theories evolved or were replaced by others over time.

These are topics very clearly and diligently discussed by Mansoor Niaz in the present book. Niaz certainly makes a good case for the inclusion of history and philosophy of science in the teaching of science, in this case chemistry. However, the most striking conclusion of the present book is not just that it is useful to take history and philosophy of science to enhance science teaching; it is that history and philosophy of science are already included in curricula and textbooks, but we often fail to make a good use of them. In the case of chemistry, when students are taught about laws, theories, and models, there are often implicit historical and philosophical elements in the respective textbook accounts. Yet, exactly because these are implicit—and they are also quite often presented in an inconsistent manner—teachers and students overlook them. Therefore, scientific concepts appear as more abstract than they should appear and often seem to come from nowhere—or in the best cases to emerge in the minds of individuals.

History and philosophy of science, in contrast, can provide a more humane portrayal of how science is done and can meaningfully connect concepts, laws, theories, and models that are often presented as disconnected. At the same time, they can also help debunk popular myths about science, for instance, by portraying scientific achievements as the products of social processes within scientific communities rather than products of "eureka" moments of solitary geniuses. Thus, the explicit reference to history and philosophy of science can help students understand why and how particular concepts were developed and later rejected, why and how scientific theories are constantly modified in the light of new empirical evidence, why and how controversy among scientists is normal and not a problem for science, and more. Paraphrasing Dobzhansky's famous dictum about the importance of the theory of evolution for biology, I could argue roughly the same about the importance of history and philosophy of science: without history and philosophy of science, science content (concepts, models, theories, laws) seems like a pile of sundry facts that make no picture as a meaningful whole.

It is for all these reasons that books like the present one by Mansoor Niaz are not only useful but should be adopted in undergraduate courses and teacher preparation programs. It is not conceivable how students may graduate from a science department without being able to describe what a scientific model is, although they have spent years studying several of them in detail. It is not conceivable that students graduate from science departments knowing Avogadro's law or the Heitler-London-Slater-Pauling valence bond theory but are unable to explain where these names come from. As Niaz shows, history of chemistry is already inside chemistry, so let's make a good use of it.

Kostas Kampourakis
Series Editor

Preface

Almost 40 years ago, when I started teaching chemistry, I had no idea of the underlying substratum of scientific knowledge in general and chemistry in particular. Chemistry was considered to be a science and hence was practiced by scientists who in turn used the "scientific method" to churn out theories and laws. The chemistry classroom was, strictly speaking, a place to tell students about the accomplishments of the great chemists who could understand and discover what was elusive to the majority of their peers. This landscape started to change for me when I saw the picture of Jean Piaget on the cover of *Journal of Chemical Education* in 1978. The obvious question that came to mind was: What was a psychologist doing in a chemistry education journal? I am still exploring the implications of the relationship between chemistry and psychology and of the necessity of history and philosophy of science. One of the most striking features of this exploration is the changing nature of chemistry (and science) and how this is based on the personal struggles, conflicts, and controversies among scientists. I still recall that in one of my first courses on the epistemology and history of chemistry, I emphasized the changing (tentative) nature of knowledge in chemistry. To my surprise, one of the students stood up and almost exclaimed: "In my high school chemistry class, I used to ask many questions, especially with respect to how and why theories develop. However, my chemistry teacher did not like me to be so inquisitive and almost admonished me in the following terms: Chemistry is a science; it never changes—so there is no need for you to be asking so many questions, as you will not change chemistry—just learn what I am trying to teach." I wonder how many classroom teachers still convey this message to their students. This book is dedicated to such students who shared their thoughts with me and provided the incentive to keep exploring the changing nature of scientific knowledge.

Learning and teaching chemistry at the interface of scientific practice and history and philosophy of science inevitably requires collaboration between educators, historians, and philosophers of science. This multidisciplinary nature of chemistry education can provide students and teachers with a rich source of how their classroom activities can be oriented toward issues that were significant in its progress.

In writing this book, I did not have any particular course in mind. This has the advantage that the book could be adopted for various types of courses, such as teaching the nature of science, introduction to the history and philosophy of science from the perspective of chemistry, and different episodes in the history of chemistry. The intended audience for this book is secondary- and university-level teachers, science teacher educators, researchers in science education, and graduate students.

Chapter 1 introduces discipline-based education research and suggests that the history of chemistry is already "inside" chemistry. Chapter 2 explores the relationship between models, theories, and laws in philosophy of science and science education. The relative importance of the different views with respect to nature of science is the subject of Chap. 3. Understanding the changing (tentative) nature of atomic models is explored in Chap. 4. Next, in Chap. 5, the role of scientific laws in the context of understanding stoichiometry is introduced. Chapter 6 facilitates understanding of chemical bonding in the context of the continuous rivalry between valence bond and molecular orbital models. An overview of recent research in different areas of chemistry is provided in Chap. 7. The transition from empiricism to historicism to naturalism and how we can go beyond are explored in Chap. 8. The contents of this book are organized around two main themes: (a) Chaps. 2, 3, and 8 deal with understanding nature of science within a history and philosophy of science perspective and can be of interest to science educators in general; and (b) Chaps. 4, 5, 6, and 7 deal with various topics of interest to chemistry educators. Readers can select the chapters that address their particular interests.

The following are some of the salient features of this book that can help readers to follow the line of argument developed in the different chapters:

1. History of chemistry is already "inside" chemistry. Different topics of the chemistry curriculum are generally presented within a historical context. However, we need to go beyond that by presenting the dynamics of scientific progress, that is, how and why theories have changed.
2. No theory can claim to provide us with an objectively true picture of the world. All theoretical formulations are tentative and eventually superseded.
3. Domain-general and domain-specific aspects of nature of science are not dichotomous but should rather be integrated to facilitate understanding.
4. Understanding of atomic models is a never-ending quest that requires imagination, creativity, and innovative techniques in the laboratory.
5. It is the context of a problem and not the laws (e.g., laws of definite and multiple proportions) that can lead us to understand how stoichiometric relations are established.
6. In order to explain the chemical bond, valence bond and molecular orbital models have continued to develop as rivals for the last 70 years and hence the recognition of the contingency thesis, namely, that the order in which events take place can be an essential factor in the acceptance of one among empirically equivalent theories.

7. Progress in science has been characterized by a *pluralism of perspectives* which led to *methodological pluralism*, and this is precisely what chemistry education needs to emphasize.
8. A comparative overview of the presentation of different topics of the chemistry curriculum in textbooks published in Brazil, Italy, Korea, Turkey, the USA, and Venezuela (these countries were included as data was available).
9. Recent views of the following historians and philosophers of science with respect to naturalism, historicism, nature of science, chemistry, and chemistry (science) education: Gerald Holton, Ronald N. Giere, Denis Phillips, Harvey Siegel, Michael Ruse, and Alan Rocke.
10. Exploration of performance expectations as suggested by the *Next Generation Science Standards* (NGSS), in the context of chemistry education.

Cumaná, Estado Sucre, Venezuela Mansoor Niaz

Niaz Endorsements

"Professor Niaz's book is most welcome, coming at a time when there is an urgently felt need to upgrade the teaching of science. The book is a huge aid for adding to the usual way—presenting science as a series of mere facts—also the necessary mandate: to show how science is done and how science, through its history and philosophy, is part of the cultural development of humanity."

Gerald Holton, Mallinckrodt Professor of Physics & Professor of History of Science, Harvard University

"In this stimulating and sophisticated blend of history of chemistry, philosophy of science, and science pedagogy, Professor Mansoor Niaz has succeeded in offering a promising new approach to the teaching of fundamental ideas in chemistry. Historians and philosophers of chemistry—and above all, chemistry *teachers*—will find this book full of valuable and highly usable new ideas."

Alan Rocke, Case Western Reserve University

"This book artfully connects chemistry and chemistry education to the human context in which chemical science is practiced and the historical and philosophical background that illuminates that practice. Mansoor Niaz deftly weaves together historical episodes in the quest for scientific knowledge with the psychology of learning and philosophical reflections on the nature of scientific knowledge and method. The result is a compelling case for historically and philosophically informed science education. Highly recommended!"

Harvey Siegel, University of Miami

"Books that analyze the philosophy and history of science in Chemistry are quite rare. *Chemistry Education and Contributions from History and Philosophy of Science* by Mansoor Niaz is one of the rare books on the history and philosophy of chemistry and their importance in teaching this science. The book goes through all the main concepts of chemistry and analyzes the historical and philosophical developments as well as their reflections in textbooks.

Closest to my heart is Chap. 6, which is devoted to the chemical bond, the glue that holds together all matter in our earth. The chapter emphasizes the revolutionary impact of the concept of the 'covalent bond' on the chemical community and the great novelty of the idea that was conceived 11 years before quantum mechanics was able to offer the mechanism of electron pairing and covalent bonding. The author goes then to describe the emergence of two rival theories that explained the nature of the chemical bond in terms of quantum mechanics; these are valence bond (VB) and molecular orbital (MO) theories. He emphasizes the importance of having rival theories and interpretations in science and its advancement. He further argues that this VB-MO rivalry is still alive and together the two conceptual frames serve as the tool kit for thinking and doing chemistry in creative manners. The author surveys chemistry textbooks in the light of the how the books preserve or not the balance between the two theories in describing various chemical phenomena. This Talmudic approach of conceptual tension is a universal characteristic of any branch of evolving wisdom. As such, Mansoor's book would be of great utility for chemistry teachers to examine how can they become more effective teachers by recognizing the importance of conceptual tension."

Sason Shaik
Saeree K. and Louis P. Fiedler Chair in Chemistry
Director, The Lise Meitner-Minerva Center for Computational Quantum Chemistry
The Hebrew University of Jerusalem, ISRAEL

Acknowledgments

Writing this book without the support of my students and colleagues in different parts of the world would have been very difficult. My institution Universidad de Oriente (Venezuela) has provided support for my research activities over the last many years. Gerald Holton (Harvard University) has been very generous in helping me to understand the role of experimentation in science and the oil drop experiment in particular. Stephen G. Brush (University of Maryland) and Michael Weisberg (University of Pennsylvania) emphasized the need to understand the origin of the ideas that led to the development of the periodic table. Alan Rocke (Case Western Reserve University) brought to my attention the importance of the law of equivalent proportions within Dalton's atomic theory. Roald Hoffmann (Cornell University, Nobel Laureate in chemistry) and Sason Shaik (the Hebrew University of Jerusalem) pointed out the never-ending rivalry between valence bond and molecular orbital models. Denis Phillips (Stanford University) and Harvey Siegel (University of Miami) explained that detailed historical reconstructions and naturalism are compatible. Ronald Giere (University of Minnesota) provided insight by explaining how and why he naturalized his philosophy of science. Michael Ruse (Florida State University) helped me to understand the role of systematicity as part of the nature of science. I am indebted to all these scholars for having responded to my queries and thus facilitated a better understanding of the chemical, historical, and philosophical landscape.

I am grateful to Kostas Kampourakis, Editor of the Springer series, *Science: Philosophy, History and Education*, for providing this opportunity to organize my views that have been developing for many years. A special word of thanks is due to Bernadette Ohmer, Publishing Editor at Springer (Dordrecht), and her staff (Marianna Pascale) for their support, coordination, and encouragement throughout the various stages of publication.

Contents

Chapter 1
Introduction

Based on a history and philosophy of science perspective, this book explores the following questions: (a) How can we motivate students for self-learning? (b) How can we encourage students to study chemistry? (c) How can we promote a more positive image of chemistry? (d) How can we convince teachers to explore new teaching strategies? (e) How can we convince curriculum designers and textbook authors to present an image that reflects how chemistry developed and keeps progressing? Are we teaching chemistry (science) as practiced by chemists (scientists), namely, as a human enterprise, at the interface of scientific practice, and history and philosophy of science (hereafter HPS)? A review of the literature shows that the traditional instruction in science classrooms and the textbooks generally ignore such issues. According to Cardellini (2013), learning chemistry requires more than simple information, "… only personal care, encouragement, and self–esteem building can help them and get them interested in chemistry" (p. 1418).

The objective of this book is to explore the relationship between chemistry content and the underlying HPS framework, in order to facilitate conceptual understanding of chemistry concepts by high school and introductory university-level students. Most chemistry courses emphasize the traditional empiricist perspective according to which experimental findings unambiguously lead to the formulation of scientific laws and theories. Actual scientific practice, however, is much more complex and involves various interactive processes, such as presuppositions of the scientist, alternative interpretations of data, controversies among scientists having similar experimental data, inconsistencies involved in the construction of a theory, and the theory-laden nature of scientific knowledge (Niaz 2010). The role of the history of ideas, and its importance for learning science, has been recognized by the National Research Council in its Next Generation Science Standards, NGSS (NRC 2013):

> Discussions involving the history of scientific and engineering ideas, of individual practitioners' contributions, and of the applications of these endeavors are important components of a science and engineering curriculum. For many students, these aspects are the pathways

M. Niaz, *Chemistry Education and Contributions from History and Philosophy of Science*, Science: Philosophy, History and Education,
DOI 10.1007/978-3-319-26248-2_1

that capture their interest in these fields and build their identities as engaged and capable learners of science and engineering (p. 249).

According to Bunce and Robinson (1997), chemistry education research (hereafter CER) can be considered as the 'third branch of our profession' covering topics such as how and why students learn, why is chemistry difficult to learn, and what facilitates effective chemistry teaching and learning (instruction and chemistry practice being the other two branches). Furthermore, these authors suggest that research should be theory based. More recently, Teo, Goh, and Yeo (2014) have characterized CER as a:

> … form of *disciplinary-based research* conducted based upon a rigorous research design that generates evidences and informs practice. This form of disciplinary- based research takes into consideration the unique *history of chemistry concept developments*, the way chemistry knowledge is constructed, and the specific skill sets and apparatuses used in the chemistry laboratory (p. 2, italics added).

The relationship between discipline-based education research (NRC 2012) and *concept development in the history of chemistry* is thought-provoking and the subject of various chapters in this book. In order to find trends in chemistry education research in the period 2004–2013, Teo et al. (2014) have conducted a content analysis based on the following six top-tiered journals: (a) chemistry education research (*Chemistry Education Research and Practice* (CERP) and *Journal of Chemical Education* (JCE)) and (b) science education research (*International Journal of Science Education* (IJSE), *Journal of Research in Science Teaching* (JRST), *Research in Science Education* (RISE), and *Science Education* (SE)). All papers included in the study were based upon empirical studies with a focus on chemistry teaching, learning, and assessments. Out of a total of 2642 articles published in IJSE, JRST, RISE, and SE, 204 (7.7 %) were related to chemistry education. Furthermore, only four of the 204 (2.0 %) articles were related to history, philosophy, and nature of chemistry. Interestingly, of the 406 empirically-based articles published in CERP and JCE, none was related to history, philosophy, and nature of chemistry. Even if one may disagree with the authors with respect to the methodology used (as empirical can be understood in different ways) for the selection of the papers and the journals, these figures clearly show the need for more chemistry education research based on a history and philosophy of science perspective.

In another study based on a wider spectrum of international education journals, it was found that:

> There is a wealth of data on student misconceptions. However, missing from the research base are investigations of pedagogical approaches that facilitate conceptual change and evidence that change has occurred. Further, it would be important to know how durable the change actually is—how long does it last? The research unequivocally demonstrates that students do not develop strong particulate models of matter nor the concepts associated with them as they progress through the curriculum. Further there is indisputable evidence that transferring conceptual understanding *of particulate behavior* from one context to another is difficult for students. Transfer of knowledge has not been a vigorous area of research in chemistry education (Towns and Kraft 2011, p. 24, italics added).

Indeed, considerable research on alternative conceptions (misconceptions) has been conducted in many parts of the world, and it is now time to go beyond our present state of knowledge and devote more attention to teaching strategies for promoting conceptual change. Again, it seems that students in most parts of the world have difficulties with the *particulate nature of matter* and hence the importance of introducing changes in how we teach about atomic structure and the role of atomic models (see Chap. 4). Furthermore, new guidelines for chemistry education research suggest the importance of novelty, impact, and influence of the findings (Towns 2013).

Discipline-Based Education Research and "Big Ideas" in Chemistry

In order to facilitate understanding and learning of chemistry, Johnstone (1991) has introduced what has come to be known as "Johnstone's triangle," based on the premise that chemistry has three central components (domains): the macroscopic, the particulate, and the symbolic. In its discipline-based education research (DBER), the National Research Council (NRC 2012) considers the "Johnstone's triangle" as a groundbreaking contribution:

> These three domains have since provided a structure for chemistry education research. Indeed, questions about what students of chemistry know, or how teachers of chemistry ought to teach, mirror the quest of chemists to connect the macroscopic properties (color, smell, taste, solubility, etc.) of matter to the structure and particulate nature of matter (p. 47).

As suggested by Johnstone, recent research in chemistry education has pursued the link between macro and micro levels of thinking and how it differs from "dexterity" in solving routine algorithmic problems. According to Tsaparlis (2014a): "If we turn however to matters of conceptual understanding, we realize that our students are as a rule ignorant and cannot answer questions such as: why chlorine appears with so many oxidation numbers, why spontaneous endothermic reactions exist, and why reactions lead in general to chemical equilibrium" (p. 42).

In order to achieve these objectives, the NRC (2012) recommends the following areas of current research: (a) students' conceptual understanding, especially of the *particulate nature of matter*, (b) use of technology to shape student reasoning, (c) analysis of student argumentation patterns, (d) use of heuristics in student reasoning, and (e) development of assessment tools to measure thinking about chemistry. The use of heuristics and argumentation patterns is particularly helpful within a history and philosophy of science perspective and is supported by current research in science and chemistry education (Niaz 2012a). Again, the importance of the *particulate nature of matter* has been recognized in recent research: "Conceptual understanding of the behavior of particles in chemical processes is very different from algorithmic or mathematical problem solving" (Williamson 2014, p. 193). (Note: Some of the recommendations of NRC are beyond the scope of this book).

Indeed, this is an ambitious agenda that can be developed and adapted according to the educational needs of students and the research backgrounds of teachers and researchers. Pienta (2012), Editor of *Journal of Chemical Education,* has endorsed these recommendations of the National Research Council with respect to the Discipline–Based Education Research: "Evidence–based practices are the most likely to produce transformative changes, ones that are persistent and will produce the most progress in teaching and learning chemistry" (p. 89).

Although, most researchers in chemistry education would agree with discipline-based education research (DBER), it still does not explicitly guide a practicing teacher or researcher. In other words, researchers in chemistry education have the daunting task of deciding how to develop teaching strategies within the context of the different topics of the chemistry curriculum. The concept of "big ideas" (AAAS 1993; NRC 1996) has been suggested as one way of implementing DBER in the classroom. For example, the statement "all matter is particulate in nature" can be considered as a big idea in chemistry (Talanquer 2013, p. 833). Despite its importance, without an adequate context, it still remains a prescriptive statement and hence students' difficulties in understanding atomic structure..

Let us consider another example of a "big idea" suggested by a distinguished chemistry educator: Chemical bonds are formed by electrostatic attractions between positively charged cores and negatively charged valence electrons (Gillespie 1997b). This is, of course, the rationale behind the valence-shell electron-pair repulsion model (VSEPR) formulated by Gillespie (Gillespie 2008; Gillespie and Nyholm 1957; also for more details, see Cardellini 2010). The VSEPR model in turn helps to extend the writing of Lewis structures for covalent bonds by including additional rules for computing bond angles. In other words, students have to memorize even more rules (for details, see Chaps. 6 and 8). General chemistry textbooks devote almost 20–25 pages with illustrations in color of the formation of Lewis structures (Niaz 2001c). However, recent research shows that despite all the efforts of textbook authors, students have considerable difficulty in drawing the structures and understanding the formation of covalent bonds. Cooper, Grove, Underwood, and Klymkowsky (2010) found that textbooks provide students with "foolproof" rules for writing the Lewis structures, and this is precisely in conflict with much of what research suggests about how students learn. In other words, if students are only provided with prescriptive formulas to memorize and solve problems with no interaction with the context to be learned, it leaves little room for conceptual understanding:

> … it became apparent that students did not appreciate the value of Lewis structures in terms of understanding why a particular molecule had particular macroscopic properties, arguably the only reason one would want students to draw them in the first place. It was clear that drawing such structures was not meaningful for these students, a fact that contributed to their difficulties with both the task and the course as a whole (Cooper and Klymkowsky 2013, p. 1119).

In a subsequent study based on their CLUE (chemistry, life, the universe, and everything) curriculum, which emphasizes learning progressions around the cross-cutting themes of structure and function, it was found that students were better at drawing and understanding Lewis structures (Cooper et al. 2012).

According to Talanquer (2013), a major difficulty in improving teaching of chemistry is perhaps the training of the teachers themselves:

> Many young, as well as some experienced, chemistry teachers seem to hold a monofaceted and unproblematic view of the subject matter they teach. These educators believe that the central challenge in teaching chemistry is identifying the best activities to engage students with prescribed textbook content, rarely questioning or critically analyzing the subject matter. Unfortunately, this lack of reflection on the content to be taught tends to thwart the most creative teaching plans (p. 832).

This clearly shows the complexity of the issues involved, as we have to improve the textbooks, motivate students, and provide appropriate feedback to teachers.

At this stage, it would be helpful to consider another example of a "big idea" suggested by Gillespie (1997a): "Putting observations first shows students that the theories and principles that are so large a part of General Chemistry are there not just to be learned, but to help in understanding these observations" (p. 485). At first sight, this seems reasonable (cf. Niaz 1999). Nevertheless, "putting observations first" requires some clarification. Philosophers of science have shown that all observations are theory-laden (see e.g. Duhem 1914; Lakatos 1970). In other words, reporting the results of an experiment necessarily involves interpretation. History of chemistry illustrates this tension between those who wanted scientific theories to be verifiable by direct experimentation and those who worked with hypothetical models. Rutherford (1915) explained this aspect of chemistry in cogent terms:

> The great majority of scientific men now regard the atomic theory not only as a working hypothesis of great value but as affording a correct description of one stage of the subdivision of matter. While this is undoubtedly the case today, it is of interest to recall that less than 20 years ago there was a revolt by a limited number of scientific men against the domination of the atomic theory in chemistry. The followers of this school considered that atomic theory should be regarded as a mere hypothesis, which was of necessity unverifiable by direct experiment, and should, therefore, not be employed as a basis of explanation of chemistry … This tendency advanced so far that textbooks of chemistry were written in which the word atom or molecule was taboo, and chemistry was based instead on the law of combination in multiple proportion (p. 176).

The reference to "revolt" in this passage refers, of course, to such distinguished scientists, such as E. Mach and W. Ostwald. Again, the use of the law of multiple proportions was one way of avoiding the atomic theory and has a long history in the development of chemistry (see Chap. 4 for details, and also this chapter). This statement from Rutherford (an experimentalist par excellence) is thought-provoking and shows how chemists did understand the scientific enterprise with considerable acumen and even foresight, with respect to the relationship between hypotheses, theories, and experimental evidence. Furthermore, students would find it interesting to note that it was written in 1915 long before philosophers of science presented their critique of positivist methodology.

The Role of History and Philosophy of Science

Continuing with the importance of the history of chemistry (atomic theory) as evidenced from the statement by Rutherford (1915), at this stage, it would be interesting to trace the origins of the role of history and philosophy of science in chemistry education (Niaz 2008, 2012a). Early in the twentieth century, positivist philosopher and physical chemist, Ostwald (1908) was perhaps the first to emphasize the importance of history and philosophy of science. Smith (1925), cofounder of the American Chemical Society's Division of the History of Chemistry, innovated by including details in courses about the personal lives of the scientists, anecdotes, crucial historic experiments and even recommended the philosophy of chemistry. Reinmuth (1932), editor of the *Journal of Chemical Education*, in an editorial, recognized the importance of the historical approach to teaching chemistry: "… it is much more important that he be shown how conclusions are reached on the basis of experimental evidence than that he memorize the conclusions. Too many students acquire the idea that scientific laws, theories and hypotheses spring full-armed from the brains of geniuses as the result of some occult phenomenon which the average man never experiences" (p. 1140). Interestingly, this comes quite close and antecedes by about 30 years Schwab's (1962) advice that science cannot be taught as "rhetoric of conclusions." Similarly, Brush (1989) has cautioned that the historical approach does not consist in merely an "assertion of the conclusions" scientists have reached in the past, but rather, "… to show how they were reached and what alternatives were plausibly advocated …" (p. 61). Indeed, Brush's advice is important if we want to avoid memorization of science content.

One of the first studies to evaluate historical materials in high school chemistry textbooks noted in the very first paragraph: "… many teachers believe that the 'historical approach' is most valuable in gaining the interest of the students. During the past ten years more than 350 historical and biographical articles have appeared in only two periodicals dealing with the teaching of chemistry" (Frank and Lundsted 1935, p. 367). The two periodicals mentioned were the *Journal of Chemical Education* and *School Science and Mathematics*. The study evaluated 20 textbooks published between 1924 and 1934 and found that the textbooks could be classified in the following categories: (a) Brief historical accounts of the development of modern chemistry, starting with alchemy, (b) short historical items scattered throughout the text, (c) short mention of the names of important persons in chemical history, and (d) biographical accounts. Dunbar (1938) evaluated 20 college general chemistry textbooks published in the period 1918 and 1937 and concluded that: (a) None of the textbooks presented a complete or consistent history of the development of chemistry. (b) Most textbooks used portraits and historical passages. (c) In several cases, numerous names and events were simply mentioned without any consistent continuity or development of the historical approach. These approaches to introducing the history of chemistry were the subject of criticism, as Jaffe (1938) noted early: "… the value of the historical material is universally recognized. Yet, the great mass of our chemistry teachers follow the syllabus and the textbook slavishly and have neither the urge nor the will to search out historical references and classical researches in an effort to vitalize their lessons" (p. 385).

Another early defense of the inclusion of HPS was provided by de Milt (1952) in cogent terms: "If it is assumed that the graduate student is not only pursuing a course of study to attain professional competence, but is also losing no opportunity to come to an understanding of the sciences related to his special field … then, it seems to me, that he can achieve this understanding by knowing something of how it all came about. This is primarily the history of science …" (p. 340). Next, de Milt asked a very pertinent question, which has also been raised by many of our students, "But why bring in the philosophy of science?" (p. 340) and responded in the following terms: "A collection of facts does not constitute a science; the important aspect of facts about the universe around us is the ideas that grow out of thinking about their relations to each other. As soon as man began to think about the relationship of facts he became a philosopher" (p. 340). A philosopher of science may express the same ideas in more professional terms. However, it is important to note that chemists and science educators themselves have raised the same issues in their own literature with respect to HPS. For example, according to Matthews (1988): "Philosophy is not far below the surface in any science classroom. At a most basic level, any text or scientific discussion will contain terms such as *law, theory, model, explanation, cause, truth, knowledge, hypothesis, confirmation, observation, evidence, idealization, time, space, fields,* and *species* … Philosophy begins when students and teachers slow down the science lesson and ask what the above terms mean …" (pp. 168–169, italics in the original). The crux of the issue is: Are we prepared to slow down the science lesson? Of course, at this stage, it is important to recognize the contributions of work done by Conant (1947, 1948) and Klopfer (1969) for the introduction of HPS in science education.

History of Chemistry Is "Inside" Chemistry

A curious student may wonder: if the importance of history of chemistry and philosophy of science for chemistry education was recognized as early as the first decades of the twentieth century, why are we still arguing for its inclusion in textbooks and the curricula? What is even more difficult to explain is why our present-day chemistry textbooks have neglected the historical approach? No easy answer is available for this question. Nevertheless, I would like to suggest two possible reasons:

(a) Brush (1978) has presented a similar critique and recommended that, "… the new style of history of science, which emphasizes the dynamics of scientific change and its relation to philosophical, technological, and social background, is much more suitable for educational purposes than the older tradition that stressed the accumulation of credit for discoveries" (p. 289). In a similar vein, historian of chemistry Trevor Levere (2006) has expressed the same idea in cogent terms: "… many authors of science textbooks still write as if there were such a thing as *the* scientific method, and use labels like induction, empiricism, and falsification in simplistic ways that bear little relation to science as it is practiced" (pp. 115–116, italics in the original). Indeed, "science as it is practiced" can be an important guideline for chemistry textbooks and teaching chemistry in the classroom. Holton (1969) has responded to the question by referring to the myth of *experimenticism*: "Almost every science textbook of necessity places a high value on unambiguous, inductive reasoning. The norm of rationalism in the classroom would seem to be threatened if the text were to allow that a correct inductive generalization may be made without unambiguous experimental evidence. Hence, the likelihood is *a priori* great that any pedagogic presentation of any subject will suggest a clear genetic link from experiment to theory" (p. 974, original italics). The genetic link that Holton has referred to is truly the *leitmotiv* of most general chemistry textbooks.

(b) Most of the early and even recent research in chemistry education (especially with respect to textbooks) has revealed that history of chemistry is generally presented as anecdotes, pictures of scientists in their labs, biographical accounts, or narration of facts devoid of any conflict or controversy. Furthermore, some of the courses especially designed for introducing the history of chemistry have not been well received by the students, as they consider that it is not part of the chemistry course, and hence irrelevant. On the other hand, in my research, I have tried to "weave" the history and philosophy of science into the fabric of the topic under study, based on personal vicissitudes of the scientists and the controversies surrounding the development of new ideas and theories (Niaz 2011). In other words, the history of chemistry is already "inside" chemistry, and this book shall provide various examples of such classroom experiences. Interestingly, a similar thesis has been presented in the case of physics: "We are not interested in adding the history of physics to teaching physics, as an optional subject: the history of physics is 'inside' physics" (Bevilacqua and Bordoni 1998, p. 451). This is an important thesis, which has generally been ignored in the science education literature. Similarly, Niaz and Rodríguez (2001), based on a historical framework, have shown that the HPS is already "inside" chemistry, and we do not necessarily need separate courses for its introduction.

Chemists and the History and Philosophy of Science

It is generally believed that chemistry is primarily an inductive science and, as compared to physics, does not provide any explanations of the phenomena observed. Many philosophers of science would agree with this or at least with the thesis that laws in chemistry play a different role compared to physics. According to physicist–philosopher of science J. Cushing (1991), one of the most important characteristics of research in physics is that it provides an explanation of what is observed. In this section, I will provide evidence to show that, although to a much lesser degree, chemists also make an effort to understand and explain experimental observations. Of course, chemistry as a science has to be theoretical. However, the question we are faced with is the degree to which the role of theoretical formulations in chemistry and physics is the same.

Mendeleev and the Periodic Table

Mendeleev's periodic table has been the subject of considerable research in the history of science, philosophy of science, and science education literature (Brush 1996; Bensaude-Vincent 1986; Brito et al. 2005; Shapere 1977; Lipton 2005a, b; Niaz et al. 2004; Niaz and Luiggi 2014). Many scholars attribute the success of the periodic law primarily to empirically observed properties of the chemical elements (inductive generalization). Many chemistry textbooks give the impression that Mendeleev had no theory or framework to support his periodic table. Based on a historical reconstruction, Niaz et al. (2004) have shown that despite some ambiguity and ambivalence, Mendeleev did have a theoretical framework, primarily Dalton's atomic theory, to provide an explanation of the periodic law. As evidence for this, I will first present Mendeleev's own writings and then those of some modern philosophers of science.

In his famous Faraday Lecture, Mendeleev (1889) stated: "The periodic law has clearly shown that the masses of the atoms increase abruptly, by steps, which are clearly connected in some way with Dalton's law of multiple proportions … While connecting by new bonds the theory of the chemical elements with *Dalton's theory of multiple proportions, or atomic structures of bodies*, the periodic law opened for natural philosophy a new and wide field for speculation" (p. 642, emphasis added). Interestingly, Mendeleev even seems to be considering the law of multiple proportions as synonymous with Dalton's atomic theory. In order to leave no doubt, in the same Faraday Lecture, Mendeleev (1889) not only attributed the success of the periodic law to Cannizaro's ideas on the atomic theory (pp. 636–637) but also went on to explicitly formulate the following hypothesis: "The veil which conceals the true conception of mass, it nevertheless indicated that the explanation of that conception must be searched for in the masses of atoms; the more so, *as all masses are nothing but aggregations, or additions, of chemical atoms*" (p. 640, emphasis added). Apparently, Mendeleev's dilemma was that, on the one hand, he could rightly claim that the periodic law was based on experimental properties of the elements (an aspiration of scientists in the late nineteenth century, that is social and historic milieu) and, on the other hand, he could not give up the bigger challenge, namely, the possible causes of periodicity, and hence the importance of the atomic theory as an explanation. Cooper (1992) has referred to this as the *milieu of the time*: "A question often very puzzling to students is why such a thing was done at such a time. Frequently the answer can only be given in the *milieu of the time*—the problems that seemed important, the opinions of the people involved" (p. xii, Preface, emphasis added).

Although, there is no consensus among philosophers of science with respect to the theoretical status of Mendeleev's periodic law, there are some who have endorsed Mendeleev's own understanding with respect to the underlying cause of periodicity. Wartofsky (1968) has clearly pointed out the hypothesis underlying Mendeleev's framework: "Mendeleev, for example, predicted that the blank space of atomic number 32, which lies between silicon and tin in the vertical column, would contain an element which was grayish-white, would be unaffected by acids and alkalis, and would give a white oxide when burned in air, and when he predicted also its atomic weight, atomic volume, density and boiling point, *he was using the periodic table as a hypothesis from which predictions could be deduced.* This was in 1871" (p. 203, emphasis added). Indeed, this clearly explains how Mendeleev may have reasoned in order to make predictions based on a hypothesis derived from his periodic law. In other words, Mendeleev's contribution can be considered as a theory. Weisberg (2007) has elaborated further on the reasoning employed by Mendeleev: "… Mendeleev had no empirical knowledge that there were any empty slots to be filled … He first needed to hypothesize the existence of the missing elements by analyzing the *theoretical structure he had created.* Then he was able to use the trends posited by the Periodic Table to make predictions about the properties of the 'missing' elements. This prediction was a theoretical, not merely classificatory, achievement" (p. 214, emphasis added). The distinction between classification and prediction is important. Those philosophers of science who deny that Mendeleev's contribution was a theory also assert that it was merely a *classificatory scheme.* Actually, Mendeleev not only classified the elements into groups and periods but also predicted the existence of "missing" elements, and this is precisely what a theory does. Gillespie (1997b) has recognized this distinction and acknowledged that Mendeleev elaborated the periodic table, "… long before anything was known about the detailed structure of the atom, as a means for *classifying and better understanding* the properties of the elements and their compounds" (p. 863, italics added).

Again, it has been argued that Mendeleev's contribution had many deficiencies and some of the predictions he made were not borne out. Interestingly, this is precisely yet another facet of all scientific theories, namely, that changes are always necessary to make the predictive power of the theory more in accord with experimental observations. Weisberg (2007) has once again explained this aspect of Mendeleev's theoretical framework: "While it is true that Mendeleev's periodic system is in need of further theoretical explanation, the same could be said of any theory that is not a fundamental physical one" (p. 215).

These arguments and counterarguments in the context of Mendeleev's periodic table are quite illustrative of how chemists work and contribute toward the formulation of theoretically based explanatory frameworks.

Lewis and the Covalent Bond

The explanation of chemical properties of atoms and molecules and the postulation of the covalent bond were two of the major contributions of the chemists toward the development of scientific progress. At the beginning of the twentieth century, it was difficult to conceptualize the sharing of electrons based on Coulomb's law (two electrons with the same electric charge and occupying the same space would exert repulsive forces). Rodebush (1928), a chemist, expressed this concern in cogent terms: "Since according to Coulomb's law two electrons should exert a repulsion for each other, the pairing of electrons seems at first glance to be a *bizarre idea.* In order to account for the peculiar behavior Lewis assumed the existence of a magnetic attraction between the electrons" (pp. 513–514, italics added).

When first proposed in 1902 (it was published later in 1916), Lewis's theory of the cubic atom based on the sharing of electrons was completely out of tune with the established belief in which the paradigm was the ionic bond. The genesis of the cubic atom can be traced to an unpublished memorandum written by Lewis in 1902 and recounted by him in the following terms:

> In the year 1902 (while I was attempting to explain to an elementary class in chemistry some of the ideas involved in the periodic law) becoming interested in the new theory of the electron (Thomson's discovery of the electron in 1897), and combining this idea with those which are implied in the periodic classification, I formed *an idea of the inner structure of the atom* (model of the cubic atom) which, although it contained crudities, I have ever since regarded as representing essentially the arrangement of the electrons in the atom. (Lewis 1923, pp. 29–30, emphasis added)

Arabatzis (2006) has questioned that the discovery of the electron should be attributed to J.J. Thomson. Postulation of the cubic atom by Lewis was essentially a "speculative" idea and was considered controversial due to the hegemony of the ionic bond based on the influential ideas of J.J. Thomson. Despite this, the cubic atom turned out to be fundamental in understanding the covalent bond and helped in understanding Pauli's exclusion principle postulated much later in 1925. Again, it is not surprising that most general chemistry textbooks ignore this important contribution toward understanding progress in chemistry (for details, see Niaz 2001c).

Pauling and the Atomic Theory

The law of multiple proportions is generally attributed to John Dalton, who first discovered it in August 1803 while working on the composition of the hydrocarbons. In the early nineteenth century, based on the inductivist versions of the history of science, it was suggested that Dalton was led to his atomic theory by the discovery of the law of multiple proportions (cf. Niaz 2012a, p. 18). The dean of modern chemistry, Linus Pauling (1964) has clarified the controversy by declaring categorically: "The discovery of the law of simple multiple proportions was the first great success of Dalton's atomic theory. This law was not induced from experimental results, but was derived from the theory, and then tested by experiments" (p. 26). Interestingly, Pauling stated this in his textbook of general chemistry. The origin and the role of Dalton's atomic theory are discussed in Chap. 4 of this book.

These three examples (Mendeleev, Lewis, and Pauling) clearly show that starting from the early-nineteenth century chemists have devoted themselves to not only doing experiments but also to providing explanations, and like all theoretical frameworks, these were tentative and needed improvements. Furthermore, it is interesting to note that some of the original ideas of Mendeleev, Lewis, and Pauling were developed while they were preparing materials for their introductory-level chemistry courses. Indeed, Gallego Badillo et al. (2012) consider that chemistry is a science that was "constructed" in the classroom. More recently, Roald Hoffmann (2012), Nobel Laureate in chemistry, has stressed: "Teaching and research are inseparable. The struggle to do both well enriches our personal intellectual lives, and enhances our contributions to society" (p. 296). In other words, these scientists clarified their own thinking of concepts, laws, and theories while preparing to help their students' understanding. This provides an opportunity to understand how history of chemistry is "inside" chemistry, and consequently the history and philosophy of science can provide us with guidance both in writing textbooks and in the classroom.

This sets the stage to understand that history and philosophy of science have been of considerable interest and guidance to chemists. Recent research has also drawn attention to the importance of philosophy of chemistry (Weisberg et al. 2011). This leads to the question: Should we follow philosophy of chemistry or the application of history and philosophy of science to understand progress in chemistry (for further details, see Niaz 2012b)? This book has followed the second approach, without of course denying the importance of philosophy of chemistry. Bensaude-Vincent (2014), a leading scholar in the history of chemistry and philosophy of science, has provided sound advice:

> This paper advocates an integration of chemistry into the philosophy of science, but it does not encourage a disciplinary partition of the field. Chemistry deserves more philosophical attention not so much to do justice to a long-neglected science or to enhance the cultural prestige of chemistry, but to undermine a number of taken-for-granted assumptions about scientific rationality and more importantly to diversify our metaphysical views of nature and reality. Because over the course of many centuries chemists have developed a special access to nature through the laboratory and a special way of investigating and dealing with material substances, they have confronted a number of epistemological and ontological issues that are worth discussing. In brief, *this paper does not make the case for a philosophy of chemistry*. It rather urges philosophers of science to listen to chemists and pay attention to what they can learn from them (p. 2, italics added, published in *Hyle*, a journal that espouses philosophy of chemistry).

This leads us to understand: philosophy *of* chemistry or philosophy *with* chemistry, namely, chemistry, deserves (not necessarily philosophy of chemistry) more philosophical attention, and one way to do that is to understand chemistry within a philosophical perspective.

Finally, application of HPS to chemistry education has become a robust area of research, and according to Erduran (2013): "The scholarship in the area is ripe for further studies. The fundamental questions such as 'What is chemical knowledge and how does it develop? What criteria, standards and heuristics shape its development?' are directly relevant for ensuring that teaching and learning environments are effectively structured and resourced for sound and deep understanding of chemistry" (p. 1561). Similarly, the need for research in chemistry education based on HPS has been recognized by various research groups in different parts of the world (Abd-El-Khalick 2005; De Berg 2014a, b; Erduran and Mugaloglu 2014; Garritz 2010; Höttecke et al. 2012; Labarca et al. 2013; Vesterinen and Aksela 2013).

Chapter Outlines

The objective of the chapter outlines is to provide the reader an overview of the chapter by including some salient features. Some outlines are longer, due to the length of the chapter.

Models, Theories, and Laws in Philosophy of Science and Science Education (Chap. 2) Models, theories, and laws are important in both philosophy of science and science education. However, there is a continuous discussion among philosophers of science about these concepts, and it is difficult to achieve consensus. Similarly, in science education, there is a continuous debate that leads to controversies. In the face of these difficulties, some scholars have suggested that we follow "science as practiced by scientists" (Levere 2006). History of science, however, shows that on many occasions what the scientists publish in their original papers is quite different from what they actually did. Given the complexity of the issues involved, some science educators have suggested that after having adopted empiricism and historicism in the past, science educators now need to adopt the model-based view based on naturalized philosophy of science (Duschl and Grandy 2013). The model-based view seems to emphasize cognitive psychology and ignore the historical reconstructions. Ronald Giere's naturalism provided one possible alternative by placing the philosophy of science at the same level as history of science, referred to as perspectivism (Giere 2006a, b). Other philosophers (Denis Phillips and Harvey Siegel) consider that the historical reconstructions can even extend the naturalistic philosophy of science. One way to retain the history of science in the curriculum is to follow the tactics and strategies of the eighteenth- and nineteenth-century scientists as suggested by James Conant. In this context, I have presented evidence of various episodes from the twentieth-century science that can also qualify for inclusion in the science curriculum, such as the oil drop experiment, the alpha-particle experiments, the atomic models and the ensuing controversies, Millikan's determination of Planck's constant and rejection of Einstein's quantum hypothesis, the wave–particle duality, and Martin Perl's discovery of the tau lepton. An important underlying aspect of these historical episodes is that scientific knowledge is perspectival rather than absolutely objective, leading to the tentative nature of scientific theories. Furthermore, besides underdetermination, some contingency is always present in any science, that is, the same experimental observations can be explained by rival theories, and their acceptance may depend on the order in which these are presented to the scientific community (e.g., quantum mechanics and bond formation).

The Nature of Science in Science Education: An Integrated View (Chap. 3) What is science and how it progresses is an important part of understanding the nature of science. Science educators have shown interest in understanding the nature of science, and there is considerable controversy with respect to its conceptualization and implementation in the classroom. Some science educators consider that there are two ways of understanding the nature of science: (a) Domain-general (based on explicit references to the following consensus-based heuristic principles: empirical nature of science, competition among rival theories, different interpretations of the same experimental data, theory-laden nature of observations, tentative nature of scientific knowledge, social and historical milieu, and others) and (b) domain-specific (cognitive, epistemic, and social practices, such as model building, observing, arguing from evidence and explaining based on a specific context of the science curriculum). I have argued in this chapter that domain-general heuristic principles are themselves derived by philosophers of science from an in-depth, domain-specific historical reconstruction of particular episodes. Consequently, understanding the nature of science as domain-general or domain-specific is a false dichotomy. Instead, I have shown with various examples from the history of science that both the domain-general and the domain-specific aspects of the nature of science complement each other, and hence, we need an integrated view. For example, if the Michelson–Morley experiment had been done at the time of Copernicus (one of the reviewers had some reservations with respect to this example, and hence, it is important to note that this is an hypothetical scenario) and not in the late nineteenth century, its result would have had no significance for the astronomers, as they considered the earth to stand still and at the center of the universe. Consequently, the historical and social milieu is an important aspect of the nature of science if we want students to understand how the ideas evolved. A study designed to introduce the integrated view of the nature of science to in-service teachers is also presented in this chapter. Results obtained show that given the necessary experience (domain-specific historical episodes), in-service teachers are quite receptive and willing to give up some of their well-ingrained aspects of an empiricist epistemology. Studies relating to the following aspects are included: (a) presentation of the nature of science in chemistry textbooks, (b) students' and teachers' understanding of the nature of science, and (c) teaching the nature of science in the classroom. Various aspects of the relationship between the nature of science and the Next Generation Science Standards, NGSS (NRC 2013) are discussed.

Understanding Atomic Models in Chemistry: Why Do Models Change? (Chap. 4) Understanding the role of early Greek philosophers (Democritus) and J. Dalton in developing the atomic theory is controversial among historians and philosophers of science. A. Chalmers (2009) claims that Dalton's theory had no testable content. On the contrary, A. Rocke (2013a) considers that Dalton's atomism is a successful theory. A study designed to evaluate the presentation of Dalton's atomic theory in general chemistry textbooks (published in USA) revealed that most textbooks stated that the atomic vision of Democritus was based on hypothetical questions (thought experiments), whereas Dalton based his theory on reproducible experimental results. Another study designed to evaluate the presentation of the atomic models of J.J. Thomson, E. Rutherford, and N. Bohr in general chemistry textbooks (published in USA) revealed that most textbooks lack a historical perspective (although historical models are being presented) and provide a simplistic view of scientific models and how these change with no reference to the difficulties and controversies involved. Exactly the same HPS-based criteria were also used to evaluate textbooks published in Turkey, Venezuela, and Korea (general physics). The similarities of the textbooks published in four countries with different cultures and languages suggest that these textbooks have an underlying common thread, namely, the dominant empiricist epistemology. Due to the difficulties faced by Bohr's model, A. Sommerfeld postulated elliptical orbits that provided greater stability to the atoms, leading to the Bohr–Sommerfeld model. A study designed to evaluate the presentation of the Bohr–Sommerfeld model in general chemistry textbooks (published in Italy and the USA) revealed that very few presented this model satisfactorily. Once again, textbooks published in two different cultures and languages were found to be very similar. Despite its success, the Bohr–Sommerfeld model went no further than the alkali metals, which led scientists to look for other models. These difficulties were resolved by the Pauli's exclusion principle and the wave mechanical model of the atom. It is concluded that understanding of atomic structure is a never-ending quest that requires imagination, creativity, and innovative techniques in the laboratory.

Understanding Stoichiometry: Do Scientific Laws Help in Learning Science? (Chap. 5) Stoichiometry is considered a difficult topic for students as understanding depends on various other topics, such as the particulate nature of matter, the concept of mole, Avogadro's number, conservation of matter, balancing chemical equations, and the laws of definite and multiple proportions. Furthermore, according to A. Rocke, from the historical perspective, laws of definite and multiple proportions are nothing more than special cases of the law of equivalent proportions. This chapter reports the design of a teaching strategy based on a history and philosophy of science framework to facilitate high school students' understanding of stoichiometry. Control group students received instruction in which the laws of definite and multiple proportions were defined as definitive and irrefutable and applied as algorithms. Experimental group students used a dialectic constructivist strategy based on the presentation of hypothetical experimental data, leading to cognitive conflicts and to a critical confrontation of different propositions. Based on the HPS framework (Giere and others), the instructor avoided defining the laws of definite and multiple proportions, unless the students themselves used these terms. Based on a posttest, results obtained revealed that experimental group students performed better than those in the control group, not only on the algorithmic problems but also on problems requiring conceptual understanding. It is concluded that if scientific laws are idealizations, then they do not describe the behavior of actual bodies, and hence may not be very helpful in understanding the empirical world.

Understanding Valence Bond and Molecular Orbital Models: Contingency at Work (Chap. 6) A historical reconstruction of the origin of the covalent bond shows that the idea of sharing electrons (covalent bond) posed considerable constraints for the scientists. When G.N. Lewis first introduced the idea of sharing electrons in 1916 based on his cubic atom, it was completely out of tune with established belief and was considered as bizarre and absurd. Pauli's exclusion principle was the next step in understanding the covalent bond. Later in the 1930s, these ideas were developed further by L. Pauling, in his valence bond (VB) model. Just at about the same time a rival model (molecular orbital, MO) was developed by R. Mulliken. Both VB and MO models have drawn upon quantum mechanics and have been rivals in their explanation of the chemical bond. However, the VB model emphasizes the pictorial aspects of the model and helps visualization, whereas the MO model is more mathematical and complex. According to the contingency thesis, both models have theoretical and empirical backing and provide alternative interpretations (Gavroglu and Simões 2012). Even today, the rivalry between the two models continues (Hoffmann et al. 2003). The objective of this study is to evaluate the degree to which general chemistry textbooks recognize the importance of the contingency thesis in the development of the two models. Results obtained show that a majority of the textbooks evaluated do present both models. However, the rivalry between the two models is ignored. This study shows that there is much to learn about how atoms combine in molecules and that different theories can be used to explain the same concept. Lack of a definitive solution to a research question shows that different theories combine to provide the "truth," thus providing students with an opportunity to do further research. Discussion of the VB and MO models illustrates that no theory can provide a complete and literally correct picture of the world (Giere 2006a, b). Furthermore, the transition from Lewis's cubic atom to Pauli's exclusion principle to the valence bond model (Pauling) to the molecular orbital model (Mulliken) to what's next clearly shows the tentative nature of our understanding of valence.

An Overview of Research in Chemistry Education (Chap. 7) This chapter provides an overview of research in various topics in chemistry curricula, based on a history and philosophy of science perspective (HPS). Six criteria, based on HPS framework, were developed for analyzing the kinetic molecular theory of gases in general chemistry textbooks published in the USA. The same criteria were used to analyze general chemistry textbooks published in Turkey. Textbooks published in both countries do not provide a satisfactory description of the following aspects: the inconsistent nature of Maxwell's research program, the kinetic theory and chemical thermodynamics as rival research programs, and the difference between algorithmic/computational and conceptual problems.

Seven criteria, based on HPS framework, were developed for analyzing the periodic table of chemical elements in general chemistry textbooks published in the USA. These criteria were based on a historical reconstruction of the topic (domain-specific), and the same criteria were used to analyze high school chemistry textbooks published in Brazil. Similar to the textbooks published in the USA, the Brazilian textbooks did fairly well on the first two criteria that dealt with the empirical aspects of accommodation and prediction. None of the Brazilian textbooks provided a satisfactory description of criteria related to predictions, explanation of periodicity, and the nature of Mendeleev's contribution, again quite similar to USA textbooks. A teaching strategy to improve precisely these aspects related to the HPS of the periodic table was designed for introductory university-level students in Venezuela. Results obtained revealed that students' understanding of these aspects improved considerably after the teaching intervention.

A historical reconstruction of the origin of the covalent bond shows that when it was first proposed, it posed considerable conceptual difficulties even for scientists. Four criteria were developed to analyze general chemistry textbooks published in the USA. The same criteria were used to analyze general chemistry textbooks published in Turkey, and none of the textbooks described satisfactorily neither the role played by Lewis's cubic atom nor that the covalent bond model (sharing of electrons) had to compete with the ionic-bond model (transfer of electrons). Similar results were found for textbooks published in the USA.

Determination of the elementary electrical charge is an important part of the science curriculum in many parts of the world. Based on a historical reconstruction (domain-specific), six criteria were developed for evaluating the oil drop experiment in general chemistry textbooks published in the USA and Turkey. The presentations of this topic in textbooks in both countries are quite similar, and none of them referred to the Millikan–Ehrenhaft controversy. Another study revealed that teaching about the experiment in the laboratory continues to be difficult even with a modern apparatus.

Explanation of osmotic pressure is an important part of electrolyte solution chemistry and has been the subject of considerable controversy. Hydrationists explained the increase in osmotic pressure by an increase in the number of free water molecules that are bounded to the salt. Ionists explained the same phenomenon due to the enhanced dissociation of the salt in water. It has been suggested that this topic can facilitate a debate in the classroom based on the views of the hydrationists and the ionists.

The photoelectric effect is generally the starting point for introducing quantum theory. Based on a historical reconstruction (domain-specific), six criteria were developed for analyzing general physics and chemistry textbooks published in the USA. Presentations of both sets of textbooks are quite similar, and the majority of them ignored one of the most important aspects, namely, that R. Millikan used the Einstein equation to determine the value of Planck's constant and still rejected the underlying theory. Similarly, this aspect is ignored in laboratory manuals. A teaching strategy based on the historical aspects showed that discussions in the classroom could help students to improve their understanding.

Most textbooks introduce the concept of wave–particle duality by posing the question: if light can have both wave and particle properties, then why do particles of matter cannot also have both properties? Based on a historical reconstruction (domain-specific), six criteria were developed for analyzing general chemistry textbooks published in the USA. In general, the textbooks ignored how the concept of wave–particle duality originated and its controversial nature. Another study has suggested that the historical background of wave–particle duality can facilitate chemistry teachers' pedagogical content knowledge.

All material included in this book that was not originally in English was translated by the author along with colleagues participating in the study.

Chapter 2
Models, Theories, and Laws in Philosophy of Science and Science Education

Introduction

In his presidential address to the 2002 Biennial Meeting of the Philosophy of Science Association, J. Earman recognized the following with respect to some fundamental concepts in the philosophy of science:

> A cursory survey of the recent literature reveals the following oppositions (among others): there are no laws versus there are/must be laws; laws express relations among universals versus laws do not express such relations; laws are not/cannot be Humean supervenient versus laws are/must be Humean supervenient; laws do not/cannot contain *ceteris paribus* clauses versus laws do/must contain *ceteris paribus* clauses (Earman 2004, p. 1228).

At first sight, this may seem disconcerting to most science educators. Despite his expressed sympathy with the milder form of the "no-laws" view, Earman (2004) considers this situation to be a "disarray" rather than "disagreement" and concluded: "It is hard to imagine how there could be more disagreement about the fundamentals of the concept of laws of nature—or any other concept so basic to the philosophy of science—than currently exists in philosophy" (p. 1228). To the relief of some science educators, Earman (2004) suggests a plausible solution to our predicament: "Rather than coming at the topic of laws of physics with preconceptions … historians and philosophers of science would do better to investigate how physicists use the concept of law" (p. 1229). This sounds as a good advice as it suggests that we need to understand science as practiced by scientists (cf. Niaz 2010). Of course, understanding what scientists do can vary with respect to what a scholar may consider as being progress in science. Actually, Earman suggests that what physicists refer to by the laws of physics are a set of true principles that form a strong but simple and unified system that in turn can be used to predict and explain. As an elaboration of the "true principles," Earman provides the following statement from a physics Nobel Laureate: "Our job as physicists is to see things simply, to understand

M. Niaz, *Chemistry Education and Contributions from History and Philosophy of Science*, Science: Philosophy, History and Education,
DOI 10.1007/978-3-319-26248-2_2

a great many complicated phenomena in a unified way, in terms of a few simple principles" (Weinberg 1980, p. 515).

Before going to the next section, it would be helpful to consider one of the views ("no laws" as referred to by Earman 2004) that purports to understand science without the laws of nature, as presented by Giere (1999). This can have implications for teaching science, and these are explored in the context of stoichiometry (Chap. 4).

Science as Practiced by Scientists

How scientists do and report science is problematic for historians, philosophers of science, and science educators. What and how scientists report their research in professional journals can of course be of considerable help for understanding science. Nevertheless, Peter Medawar (1967), Nobel Laureate in medicine, has referred to the pitfalls involved in the following terms:

> It is no use looking to scientific 'papers' for they not merely conceal but actively misrepresent the reasoning that goes into the work they describe. If scientific papers are to be accepted for publication, they must be written in the inductive style. The spirit of John Stuart Mill glares out of the eyes of every editor of a Learned Journal … Only unstudied evidence will do—and that means listening at a keyhole. (p. 151)

More recently, Medawar's contribution to understanding science has been recognized positively in the following terms:

> To those who considered that the idea that science was conjectural in character in some way diminished it and, by implication, those who practiced it, Medawar retorted that nothing could be more diminishing than the idea that the scientist was the collector and classifier of fact: 'a man who cranks some well-oiled machine of discovery.' (Calver 2013, p. 308)

In many parts of the world, science education still inculcates the view that scientists know how to handle a "well-oiled machine of discovery." Medawar's critique has been endorsed by Holton (1978b), who suggested that besides the "papers," scholars can study:

> Letters, autobiographical reports cross-checked by other documents, oral-history interviews conducted by trained historians, transcripts of conversations that took place in the heat of battle at scientific meetings, and, above all, *laboratory notebooks*—firsthand documents directly rooted in the act of doing science, with all the smudges, thumbprints, and bloodstains of the personal struggle of ideas (p. 25, emphasis added).

Holton (1978a, b) himself provided the lead by reconstructing the controversy between R. Millikan (1868–1953) and F. Ehrenhaft (1879–1952), with respect to the determination of the elementary electrical charge, by consulting Millikan's handwritten notebooks at CALTECH. Acceptance of the elementary electrical charge was preceded by a bitter dispute between Millikan (University of Chicago) and Ehrenhaft (University of Vienna) that lasted for many years (1910–1925). According to the 'papers' published in prestigious journals, both Millikan and Ehrenhaft obtained quite similar results, and yet Millikan was led to formulate the elementary electrical charge (electron) and Ehrenhaft to fractional charges (sub-electron). Although Millikan was awarded the Nobel Prize in 1923, Ehrenhaft (1941) continued to critique Millikan, and for most of the scientific community, the controversy remained a mystery.

It was Holton's (1978a, b) seminal work, published almost 55 years after Millikan was awarded the Nobel Prize, which provided an insight into and an understanding of the research methodologies of the two protagonists. Holton's account based on a historical "reconstruction" of the research methodology and Millikan's handwritten notebooks revealed that there were 140 experiments on an equal number of oil drops. In the actual publication (Millikan 1913), complete data on 58 drops is meticulously presented, and furthermore Millikan emphasized that: "… *this is not a selected group of drops but represents all of the drops experimented upon during 60 consecutive days*" (Millikan 1913, p. 138, italics in the original). How do we interpret this information? The laboratory notebooks tell us that there were 140 drops, and the published "paper" is emphatic that there were 58 drops. What happened to the other 82 (59 %) drops? In other words, Millikan apparently excluded drops that did not have charge equal to an integral multiple of the elementary electrical charge. Holton (1978a) wondered about Ehrenhaft's response if he had had access to Millikan's notebook: "If Ehrenhaft had obtained such data, he would probably not have neglected the second observations [that did not give the expected value of the elementary electrical charge] and many others like it in these two notebooks that shared the same fate; he would very likely have used them all" (pp. 209–210).

Teaching science as practiced by scientists can indeed be an important objective of science education (Niaz 2010). In a recent study, Hodson and Wong (2014) have made a strong defense of the need for contacts between students and practicing scientists:

> However, the key point we wish to make in this paper is that student understanding of the complexity and diversity of scientific practice would be immeasurably helped by experiences at 'the horse's mouth', that is, by contact with practising scientists. For example, students can learn a great deal about the language, theories, methods, history, traditions and values of science by talking to scientists; listening to their stories; reading their publications; attending lectures, seminars and discussions involving scientists (learning from scientists); observing, interviewing and/or working alongside them (learning with scientists) and from what they read in well-designed case studies, Internet websites, biographical material, newspapers and the accounts of respected commentators on contemporary scientific practice (learning about scientists). (p. 2662)

Indeed, at first sight, this sounds quite reasonable and even perhaps unproblematic. However, a closer look shows that there are many caveats involved, as can be observed from a historical reconstruction of various episodes in the history of science, such as the oil drop experiment, the photoelectric effect, the wave–particle duality, the periodic table, the atomic theory, the valence bond, and the molecular orbital theories. Again, the thinking of Weinberg (2001) in physics has been critiqued by Giere (2006a, b) in cogent terms that can provide students with an opportunity to scrutinize views that come out of the "horse's mouth."

This clearly shows that following how scientists do science is complex and needs considerable elaboration, details, and clarification. In other words, it is precisely the historical reconstructions that provide one alternative for understanding how scientists have actually worked, beyond what has been published in the original papers. More recently, Holton (2014a) has elaborated on the importance of *how science is done*: "The squishy phrase 'understanding of science' can mean many things, but above all it must, I insist, include *knowledge* of science, plus an acquaintance with how science is done, plus a view of science as part of the cultural development of humanity" (p. 1876, italics in the original, underline added).

Beyond the Historical Turn in the Philosophy of Science

According to Duschl and Grandy (2013, p. 2114), over the last 100 years, there have been three major movements in philosophy of science, namely: (a) logical positivism/empiricism (hypothesis testing) that provide justification of scientific knowledge claims, represented by the work of Carnap, Hempel, Neurath, and Reichenbach, among others; (b) history-based view of theory development, based on paradigms, research programs/traditions, and heuristic principles, represented by the work of Kuhn, Lakatos, and Laudan, among others; (c) model-based view of cognitive and social dynamics based on naturalized philosophy of science represented by the work of Giere, Nersessian, Thagard, and Kitcher, among others. Duschl and Grandy (2013) categorize those science educators who subscribe to logical positivism and the history-based view as Version 1. Furthermore, they critique Version 1 science educators in the following terms: "Grounded in dated (logical positivism and historical turn) views that depict NOS through *heuristics* that focus on individual scientists justification of knowledge" (Duschl and Grandy 2013, p. 2125, Table 3; italics added). Interestingly, one reviewer of this book stated: "This is an inaccurate characterization of historicism. Historicists and Kuhn in particular, stressed the communal character of scientific knowledge." For the importance of domain-specific heuristics in science education, see the last section. As an alternative, these authors then recommend Version 2, "… as grounded in the 'Naturalized View of Philosophy of Science' that emerged among philosophers of science as another response to the historical turn" (p. 2112). Next, these authors espouse the model-based view based on naturalized philosophy of science, and following are three aspects that illustrate their perspective, which in my opinion can be considered as the "cognitive turn":

1. According to Duschl and Grandy (2013, p. 2112), the following quote from Carruthers et al. (2002) captures the current consensus among contemporary philosophers of science: "It became important, then, to see science, too, as a natural phenomenon, somehow recruiting a variety of natural processes and mechanisms—both cognitive and social—to achieve its results. Philosophers of science began to look, not just to history, but also to cognitive psychology in their search

for an understanding of scientific activity" (p. 4). I would agree that most philosophers of science today do agree with some form of naturalism. Actually Duschl and Grandy (2013) suggest that, "The *naturalistic turn* in philosophy of science was a response by philosophers to fill in the gaps left by the demolition by Kuhn and others of the basic tenets of logical positivism" (p. 2126, italics in the original). It seems that Duschl and Grandy do not recognize the importance of historical reconstructions and consequently consider the historical turn in the philosophy of science to be dated. Nevertheless, it is not entirely correct to suggest that naturalism originated as a reaction to the historical turn in the philosophy of science. According to Giere (1999, 2010), the origins of naturalism can be traced to the writings of early twentieth-century pragmatists, such as John Dewey, William James, and Otto Neurath within the Vienna Circle (for details, see next section, Giere's naturalism). Giere's (2010) major recommendation is to: "Characterize naturalism not as a thesis, but as a method" (p. 214), which constitutes a *methodological* turn. Again, the relationship between cognitive psychology and its philosophical underpinnings can be traced to the writings of Piaget (cf. Piaget and Garcia 1989, for a discussion and criticism of Popper, Kuhn, Lakatos, and Feyerabend). Furthermore, according to Kitchener (1986): "It became apparent to me as I delved into these French works [Piaget's] that there was an entire philosophical world there to be explored, one unknown to most English-speaking philosophers. I found this largely unexplored philosophical territory intensely interesting and close in spirit to contemporary philosophy of science" (p. viii). The importance of Piaget's philosophical thinking for science education has been explored in a series of articles by Niaz (1991, 1992, 1993, 1995c).

2. Duschl and Grandy also write that: "The advancement of the learning sciences (Sawyer 2006) and our deeper understanding of children's cognitive development (NRC 2007) has led us to recognize and seek coordination of a triad of practices—cognitive, epistemic and social—in the learning of science. The strong recommendation from *Taking Science to School* (*TSTS*) (NRC 2007) is that acquiring conceptual knowledge (e.g., content) should not be separated from learning science practices (e.g., processes). The emerging consensus is that science learning and teaching ought to be grounded in epistemological, social structures, and practices" (Duschl and Grandy 2013, p. 2113). One could, of course, agree with some of the findings of TSTS (elaborated under the chairmanship of R. Duschl). However, it also raises some philosophical issues with respect to Piagetian theory and its transition toward information processing, neo-Piagetian theories. In an interchange, Klahr (2009) asked Duschl: "While I fully agree with the TSTS position (having been a member of the committee that produced it under your able chairmanship!) that Piagetian stage theory is mistaken in its claims about a set of sequence of stages, I think your claim is too strong. It seems that the statement quoted above is a slightly disguised version of Bruner's now roundly refuted claim that 'any subject can be taught to any child at any stage in some form that is honest' (Bruner 1977). Can you give an example of a situation where a child is 'stalled' and a 'thoughtful and informed curriculum' can 'move

learners forward'?" (p. 323). Most science educators would agree that the sequence of Piagetian stages has problems. Given the considerable amount of research in science education with respect to Piagetian stages (Lawson 1985), this issue is of considerable importance. Now let us see how Duschl responded to Klahr's query: "Adey, Shayer, and Yates (2001) have demonstrated the effectiveness that designed thinking activities and group-learning sessions can have on advancing stalled domain-general reasoning skills" (Duschl and Duncan 2009, p. 324). Besides this, they also cite Chi (2005) and Clement (1998). Paradoxically, the important point is that among science educators, Michael Shayer and Philip Adey have precisely contributed considerably to popularize the constraints imposed by Piagetian stages in learning science.

3. Finally, Duschl and Grandy write that: "During the 1960s Piaget's ideas about child development and Vygotsky's ideas about sociocultural development were major agents of change. In fact, Kuhn (1962) invokes Piaget's ideas of conceptual change in *[The] Structure [of Scientific Revolutions]*. Within a decade with the emergence of cognitive information-processing psychology there would be debates about domain-general versus domain-specific modes of learning when comparing and contrasting ideas of Piaget and Vygotsky" (Duschl and Grandy 2013, p. 2110). It seems that after discarding Piagetian stage theory, these authors are now endorsing Bruner's idea that any subject can be taught to any child (as suggested by Klahr). Of particular interest is the study by Lehrer and Schauble (2012), in which 5th and 6th graders were introduced to a pond ecosystem. It is plausible to suggest that such studies, although helpful, are at best training studies (cf. Niaz 1997). Interestingly, however, the contributions of neo-Piagetian studies, based on information processing, have been completely ignored. A critique of Piagetian stage theory led to the formulation of various neo-Piagetian information-processing theories, of which the one by Pascual-Leone (1970, 1987) has been particularly popular in science education (BouJaoude et al. 2004; Lawson 1983; Niaz 1988, 1989; Niaz and Lawson 1985; Wu and Shah 2004). A basic premise of this research is that the cognitive complexity of a science task (domain-specific) can be reduced by "manipulating" the information processing demand and hence making the logical structure (domain-general) more accessible. This research continues to be a robust line of investigation (BouJaoude et al. 2004; St Clair-Thompson et al. 2012; Stamovlasis and Tsaparlis 2012; Tsaparlis 2014b; Tsaparlis et al. 1998; Yuan et al. 2006).

Duschl and Grandy (2013) clearly consider cognitive psychology more important than historical reconstructions in order to understand progress in science. Actually, they go beyond (or perhaps backwards!) by even considering the historical turn to be dated (see quote from p. 2125 cited above), and it seems that they are rather espousing a position that can easily be termed as the "cognitive turn." Given Duschl and Grandy's emphasis on naturalized philosophy of science, let us now consider how it originated and what exactly it means.

Giere's Naturalism

According to Giere (1999), a self-described naturalist: "Naturalism is a general program for all philosophy, indeed, for all of life" (p. 70). Furthermore, Giere has emphasized the internal connections between naturalism and pragmatism and considers early twentieth-century American pragmatists (e.g., John Dewey and William James) to be foremost champions of naturalism. Giere has recapitulated his difficulties with drawing normative (philosophical) conclusions from empirical (historical) premises and the consequent need for naturalizing the philosophy of science:

> In my 1973 BJPS review of a volume of Minnesota Studies on the relations between history of science and philosophy of science, I pointed out the conflict between the a priori, normative claims of the philosophy of science, which then was dominated by Logical Empiricism, and the empirical (naturalistic) claims of historians of science. How, in the spirit of Hume, I asked, could one get normative (philosophical) conclusions from empirical (historical) premises? How, as Kuhn suggested, could one use history of science to justify philosophy of science conclusions? I resolved this conflict for myself with my 1985 Phil of Science paper on naturalizing the philosophy of science. This move puts the philosophy of science on the same naturalistic level as history of science. The only legitimate normative claims are, as Laudan said, claims about effective means to desired goals, which can be judged empirically. (Giere 2014, p. 1)

This statement clearly shows the need for putting philosophy of science at the same level as the history of science. At this stage, a science educator may wonder about the possible role of historical reconstructions in teaching science. Interestingly, Laudan (1996), a naturalist, has endorsed the use of history of science and its reconstruction for science education. This issue will be the subject of discussion in the Conclusions (Chap. 8).

Following are some important aspects for understanding naturalism (Giere 1999): (a) Characterizing naturalism as a method and not a thesis, namely, *methodological naturalism* is to be preferred over *metaphysical naturalism* as it provides strategies for understanding the world (p. 70). (b) There should be reliance on the methods and results of current science (p. 72). (c) Scientific explanations need not be based on *a priori* arguments (p. 77). (d) The historical record shows that what counts as a scientific explanation changes over time, and a critical appraisal always remains open (p. 76). (e) Empirical representational relationship is not between statements and the world, but between models and the world (p. 73). (f) The fit between a model and the world may be like the fit between a map and the region it represents (p. 82). (g) Historically, development of models is a contingent matter. In other words, models of the world held at any given time might have been different if historical contingencies had been different (p. 77). An example of the historical development of contingency in the case of valence and molecular orbital models is provided in Chap. 6.

Giere (1999) considers Darwin's explanation of the origin of species as the most important exemplar for any form of naturalism:

> The historical background to Darwin's theory included a strong tradition of natural theology in which the design of nature, particularly the design of animals and humans, was taken as evidence for a supernatural designer and creator. Darwin, as I understand him, did not

> provide direct evidence *against* the existence of such a creator. Rather, his theory of evolution by natural selection provided an *alternative*, naturalistic explanation of the acknowledged facts, for example, of the fit between the functional anatomy of animals and their environment (pp. 70–71, italics in original).

According to another naturalist: "We need a historical perspective that leads us from the period during which the ideas espoused by the intelligent design-ers were widely accepted, through the episodes in which they were discarded in favor of Darwinian principles, to our present situation" (Kitcher 2007, p. 14). Furthermore, Kitcher (2011) considers epistemology to be blind without history.

Giere's Perspectivism

According to Earman (2004), what the physicists mean by the laws of physics are a set of *true principles* that can be used to predict and explain. Weinberg (2001), a physicist, has elaborated this further with respect to true principles and how scientists conceptualize progress in science:

> What drives us onward in the work of science is precisely the sense that there are *truths out there to be discovered*, truths that once discovered will form a permanent part of human knowledge. (p. 126, emphasis added)

This statement makes the task of the science educator much more difficult. If science looks for *truths*, then Newton's laws should have been the prime example. However, at the beginning of the twentieth century, Einstein's theory of special and general relativity and later quantum mechanics questioned Newton's perspective. Does this mean that Newton's laws were false or even that perhaps he was not being objective? Giere (2006a) has characterized such philosophical positions as "objectivist realism" (p. 5) and explained cogently:

> Weinberg should not need reminding that, at the end of the nineteenth century, physicists were as justified as they could possibly be in thinking that classical mechanics was objectively true. That confidence was shattered by the eventual success of relativity theory and quantum mechanics a generation later (p. 118).

Giere's rejection of "objectivist realism" is based precisely on an examination of scientific practice, which is recognized by historians, sociologists, psychologists, and science educators, as a human enterprise:

> … I wish to reject objective realism but still maintain a kind of realism, a perspectival realism, which I think better characterizes realism in science. For a perspectival realist, the strongest claims a scientist can legitimately make are of a qualified, conditional form: "According to this highly confirmed theory (or reliable instrument), the world seems to be roughly such and such." There is no way legitimately to take the further objectivist step and declare unconditionally: "This theory (or instrument) provides us with a complete and literally correct picture of the world itself." (Giere 2006a, pp. 5–6).

A Dilemma for Science Education

Given the complexity and importance of the issues involved, it seems that the science education community is facing a dilemma, with respect to historical turn and naturalist philosophy of science. Consequently, I decided to seek help from the following philosophers of science who at the same time have a keen interest in science education:

1. Denis C. Phillips, Stanford University
2. Harvey Siegel, University of Miami

Both philosophers were sent a request entitled "A dilemma for science education," which is presented in Box 2.1.

Box 2.1: A Dilemma for Science Education
I would like to seek your advice with respect to a recent article by Duschl and Grandy (2013). According to these authors, current philosophy of science has gone beyond the historical turn (Kuhn, Lakatos, Laudan) and now espouses a naturalist philosophy of science (Giere, Nersessian, Thagard, Kitcher). I would appreciate your response to the following questions:

1. Would you agree that at present most philosophers of science are naturalists? According to Siegel (2013): "Of particular note is the *naturalistic* character of contemporary philosophy of science … The editors are certainly correct that naturalism is the dominant view these days, and the subsequent chapters bear this out …" (p. 731).
2. If we go beyond the historical turn and adopt a naturalist stance, does this mean that science education no longer needs in-depth historical reconstructions of various scientific developments that form part of the science curriculum, such as Watson and Crick experiments to discover the DNA structure, Darwinian theory, Millikan's oil drop experiment, Michelson–Morley experiment, atomic structure (Thomson, Rutherford, others), bending of light in the 1919 eclipse experiments, etc. Interestingly, Phillips (2005) has recommended that educational research needs to do in-depth historical studies similar to those of Popper, Kuhn, Lakatos, Cartwright, and Galison.

References

Duschl, R. A., & Grandy, R. (2013). Two views about explicitly teaching nature of science. *Science & Education, 22*(9), 2109–2139.

Phillips, D. C. (2005). The contested nature of empirical educational research (and why philosophy of education offers little help). *Journal of Philosophy of Education, 39*, 577–597.

Siegel, H. (2013). Review of Stathis Psillos and Martin Curd (eds): *The Routledge companion to philosophy of science* (pp. 729–731), 2008. New York: Routledge. *Science & Education, 22*(3), 729–731.

Both philosophers were kind enough to respond, and their responses (excerpts from email) are reproduced here with their permission.

> *Denis C. Phillips (2014)* As you well know, the issues are complex, and hinge to some degree on what precisely one takes to be involved in the 'historical turn' and in 'naturalistic stance'. So my response is extremely sketchy! Basically I do not see that there is any incompatibility between using detailed historical cases and the naturalistic stance; indeed I see the cases as potentially furthering the naturalistic program. If (for whatever reason) one wants to get a realistic/naturalistic picture of what scientists actually DO instead of intuiting this account or adopting an account of science a priori, then one must use careful studies OF actual science as the basis for our analysis. Sometimes, however, a study that is being done on a contemporary scientific research program faces severe difficulties (access to all materials and communications, tricky power relations, etc. etc.) but a historical study may not have such limitations and hence may give a more 'empirically accurate' (sophisticated, detailed, etc) account. Popper, Lakatos, et al. are often classed as naturalistic, and they use historical cases for the reasons above. This does not guarantee of course that the analyses they arrive at are acceptable in the face of criticism.

> *Harvey Siegel (2014)* Yes, I agree. Naturalism is the dominant view. But people who proclaim it mean different things by it, and some are more 'naturalist' than others. And there are still holdouts like me who think it vastly overrated. I think I gave some indication of this in the review that you quote. I don't think naturalists are generally anti-historical-reconstruction. In fact, they might well think that the historical record is itself a part of the natural world and that historical research is itself 'naturalistic'. (It certainly isn't done a priori or from the armchair!).

My Comments

Interestingly, both Phillips and Siegel agree that naturalists are not necessarily anti-historical and that the historical reconstructions can even potentially extend the naturalistic program. Again, both emphasize that just like the historical reconstructions, naturalists are interested in what the scientists actually do and avoid a priori armchair considerations. Phillips even considers Popper and Lakatos to be naturalists and that historical reconstructions have to deal with the lack of availability of materials and "tricky power relations." Consequently, like any other human enterprise, reconstructions are also open to criticisms and revisions.

Conant and History of Science

At this stage, it can be argued that Duschl and Grandy (2013), in their Version 2, do endorse some form of history: "History of science cases holistic and complex renditions" (Table 3, p. 2125). However, in the same table, these authors consider Version 1 of science educators as based on dated views that include logical positivism and the historical turn. I wonder if many science educators these days would subscribe to logical positivism. Now, let us see how Duschl and Grandy conceptualize "holistic and complex renditions" of history of science. Their advice is to adopt Conant's tactics and strategies with a focus on a domain-specific perspective, and they provide the following quote:

> The stumbling way in which the ablest of the early scientists had to fight through thickets of enormous observation, misleading generalizations, inadequate formulations and unconscious prejudice is the story which it seem to me needs telling (Conant 1947, p. 15).

It seems that Conant preferred to include case studies of primarily eighteenth- and nineteenth-century science, which Duschl and Grandy seem to endorse. However, if we take Conant's advice critically within a historical perspective, then even twentieth-century science provided instances of: "thickets of enormous observation," "misleading generalizations," "inadequate formulations," and "unconscious prejudice."

Let us consider the following episodes in twentieth-century history of science, which are an important part of the science curriculum in almost all parts of the world:

(a) In the determination of the elementary electrical charge, R. Millikan and F. Ehrenhaft studied hundreds of oil and metal drops over at least 15 years (1908–1923) and elaborated different generalizations (some of which were inadequate), and both scientists had preconceived ideas with respect to the atomic nature of electricity (Holton 1978a; Niaz 2005).

(b) Based on his cathode-ray experiments, J.J. Thomson postulated his model of the atom in the first decade of the twentieth century. Soon after that, E. Rutherford's colleagues reported experimental results based on alpha-particle experiments (Geiger and Marsden 1909). This data suggested to Rutherford and colleagues the plausibility of the nuclear model of the atom. However, Rutherford was reluctant to publish his results for almost two years and after several repetitions and deliberations finally published his nuclear model in 1911 (Rutherford 1911). It is generally ignored that Thomson also conducted alpha-particle experiments in his own laboratory and found very similar results. However, in order to explain the data, Thomson propounded the hypothesis of compound scattering (that supported his model), and Rutherford propounded the hypothesis of single scattering that supported the nuclear model. A science student may wonder how two leading scientists could interpret the same experimental data in diametrically opposite ways. A bitter controversy ensued between the two that lasted for many years (Wilson 1983; Niaz 1998b, 2009, 2012a).

(c) Bohr's (1913) model of the atom is based among other sources on his now-famous four postulates. Did Bohr have experimental evidence for these postulates? If not, what was the warrant for their acceptance? Textbooks in almost all parts of the world reproduce the postulates, but very few make an attempt to provide an explanation of how they were developed to the possibly "bewildered" science students. Historians of science have devoted considerable time and effort to understanding this episode (Heilbron and Kuhn 1969; Lakatos 1970). However, as these pertain to the "historical turn" in the philosophy of science, their interpretations may be considered to be dated (Duschl and Grandy 2013). Consequently, let us consider the views of the following scholars:

> It is understandable that, in the excitement over its success, men overlooked a malformation in the theory's architecture; for Bohr's atom sat like a baroque tower upon the Gothic base of classical electrodynamics (Margenau 1950, p. 311).

> In 1913 Niels Bohr proposed his famous theory of the hydrogen atom. One cannot say that he resolved the problems raised by Rutherford. In a sense he crystallized the dilemma in an even more dramatic form. Focusing his attention entirely on the construction of a nuclear atom, Bohr took what principles of classical physics he needed and added several nonclassical hypotheses almost without precedent; the mélange was not consistent. But they formed a remarkably successful theory of the hydrogen atom. It would be years before it could be said that one had a consistent theory again (Cooper 1992, p. 325).

> The first assumption [postulate] is the existence of stationary states, the second is the frequency rule. Bohr regarded them as the unshakeable pillars of his theory. They were indeed more directly related to experiments than other assumptions of his theory. Until at least 1925, they remained the two basic postulates of the quantum theory, despite the vicissitudes of most other assumptions (Darrigol 2009, p. 154).

We have traced the evaluations and opinions of independent scholars (not "historical turn" philosophers of science, furthermore Cooper is a Nobel Laureate in physics), over a period of almost 60 years. Despite some reservations, the picture that emerges is the following: a theory's architecture can have malformations, a theory can approximate a mélange and some assumptions of a theory may not have experimental warrants. Surprisingly, this coincides to a fair degree with the conceptualization of "historical turn" philosophers of science (Lakatos 1970) and historians of science (Heilbron and Kuhn 1969).

(d) Robert Millikan strongly believed in the classical wave theory of light, which could explain the well-known phenomenon of interference, and hence he thought that it could also explain the photoelectric effect. In order to explain this effect, Einstein (1905) hypothesized that ordinary light behaves as though it consists of a stream of independent localized units of energy that he called *light quanta*. After working on the photoelectric effect for many years, Millikan (1916) designed an experiment to calculate Planck's constant, *h*, in Einstein's famous photoelectric equation ($\frac{1}{2}mv^2 = h\upsilon - p$), which was soon accepted by the scientific community. Interestingly, however, in the same publication, Millikan had this to say: "This hypothesis may well be called reckless first because an electromagnetic disturbance which remains localized in space seems a violation of the very conception of an electromagnetic disturbance, and second because it flies in the face of the *thoroughly accepted facts of interference*" (Millikan 1916, p. 355, italics added). Holton (1999) has referred to this dilemma in the following cogent terms: "So Millikan's (1916) paper is not at all, as we might now naturally consider it to be, an experimental proof of the quantum theory of light" (p. 232). No wonder, our students would like to know: Could Millikan determine Planck's constant, *h*, and still reject the underlying hypothesis of *light* quanta? To make matters worse, in his *Autobiography* published in 1950, at the age of 82, Millikan gave an entirely different version of this story and accepted that his data did provide an experimental proof of the quantum theory of light (Millikan 1950, pp. 101–102). Stuewer (1975, p. 88) considers this adjustment by Millikan as "shocking," considering the fact that even in 1924, in his Nobel Prize acceptance speech, Millikan still questioned Einstein's hypothesis of *light quanta*. At this stage, it would be interesting to know which of the two versions of the historical events is presented to our students in the textbooks. Niaz, Klassen, McMillan, and Metz (2010a) have reported that of 103 general physics textbooks (published in USA, between 1950 and 2008), 98 % presented the experimental details of Millikan's determination of Planck's constant (h) with no mention of Millikan's rejection of the underlying hypothesis of *light quanta*. This clearly shows that the lack of a historical perspective can deprive students not of the naturalist/historical stance, but rather of understanding the difficulties involved in doing research and perhaps what Phillips (2014) referred to as "tricky power relations" as an aspect of science as a human enterprise.

(e) Wave–particle duality plays an important part in understanding modern atomic structure. Einstein and de Broglie suggested (before 1923) the idea of wave–particle duality, before there was any conclusive experimental evidence to support it. Despite support from Einstein, de Broglie faced considerable controversy and difficulties in the scientific community. Influential schools of spectroscopists (Bohr's Copenhagen school and Sommerfeld's Munich school) not only provided considerable resistance to the novel wave–particle duality idea but also questioned de Broglie's reputation as a scientist. Considerable amount of experimental evidence (thickets of enormous observation) based on the work of Davisson, Germer, Thomson, and Reid became available in 1927; however, it was difficult to interpret it. Niaz and Marcano (2012) have reported that of 128 general chemistry textbooks (all published in the USA, between 1954 and 2011), almost 90 % did not mention the controversial nature of wave–particle duality and the difficulties involved in its experimental determination and acceptance by the scientific community. This clearly shows, once again, how textbooks present a "sanitized" version of the historical events and deprive students of what real science is all about (e.g., controversies, tricky power relations).

(f) Martin Perl was the recipient of the 1995 Nobel Prize in physics for his discovery of the tau lepton (elementary particle physics), based on a 16-year struggle (1963–1979), when all experimental measurements agreed with the hypothesis that the tau was a lepton produced by a known electromagnetic interaction. This hypothesis, although difficult to substantiate with experimental evidence, was formulated by Perl as early as 1963, and later he recalled this in the following terms: "I dreamed that if we could find a new lepton, the properties of the new lepton might teach us the secret of the electron-muon puzzle" "(Perl 2004, p. 407). Even in 1975, when Perl and colleagues had sufficient experimental evidence to support their hypothesis, they preferred to wait for the scientific community to obtain similar results and in the interim denoted the new particle by U for "unknown." Finally, Perl attributed his success to:

> *I had smart, resourceful, and patient research companions. I think these are the elements that should be present in speculative experimental work: a broad general plan, specific research methods, new technology, and first-class research companions. Of course, the element of luck will in the end be dominant* (Perl 2004, pp. 418–419, italics in the original, underline added).

I am sure Conant would have liked this story and not only for the reasons he stated publicly, but rather for the perseverance, dedication, supporting difficulties, humbleness in success, and above all the respect for the scientific community (for details, see Niaz 2012a, Chap. 7). Such historical reconstructions may be questioned by naturalists, but they do bring to the forefront the idea that "science in the making" is above all a human enterprise, and studying its details can be of immense help to our students. This is all the more important as most of our science textbooks distort the history of science in order to present a vision of science that still resembles logical positivism. Gooday, Lynch, Wilson, and Barsky (2008) have advocated engagement with the history of science (especially in textbooks) in order to teach science and train scientists:

> The key role of history here is characterizing the complexities of how science *changes*. So many science textbooks unhelpfully—and above all inaccurately—cultivate a rather static image of scientific disciplines, as if they were completed with comprehensive certainty. It is perhaps not difficult to understand how this gross oversimplification might arise as the result of a pedagogical need to 'tidy up' the presentation of science to meet the needs and capacities of students. But faced with the textbook spectacle of such an apparently unalterable monolith, is it any wonder that students can have difficulty conceiving how they might ever contribute to science? (p. 326, italics in the original, interestingly Wilson is a Nobel-Laureate in physics).

This section helps us to reflect upon and consider the following important issues: (a) How many reform documents published in various parts of the world consider the importance of teaching science in order to stimulate students to contribute to our existing knowledge of science? (b) Do we present a false image of science in our textbooks and classrooms? (c) Is the false image of science conducive toward a better understanding of science?

In a recent review (Niaz 2014a) of 52 studies which analyzed textbooks based on history and philosophy of science (HPS) criteria, published in major science education journals (*International Journal of Science Education, Journal of Research in Science Teaching, Science Education,* and *Science & Education*), over a 15-year period (1996–2010), it was found that most biology, chemistry, physics, and school science textbooks lack an HPS perspective.

According to Galison (2008): "Back in the postwar period, James Bryant Conant hoped that the Case Studies in Experimental Science that he organized would, by a kind of Baconian generalization, lead to a general understanding of scientific method. But it is hard to see this Baconianism emerging from microhistories today. Microhistory is supposed to be exemplification, a display through *particular* detail of something *general*, something more than itself. It is supposed to elicit the subtle interconnections of procedures, values, and symbols that mark science in a place and time, not as a method but more as a kind of scientific culture" (p. 120, italics in the original). In my opinion, science educators would be interested in microhistories but not in Baconian generalizations.

An Overview for Science Educators

Here is a summary of the questions, issues, and subjects discussed in this chapter:

1. Historical reconstructions of different episodes constitute an important means to understanding how science develops and progresses (Giere 1999, 2006a, b; Siegel 2014; Kitcher 2007; Phillips 2014). Such reconstructions provide students with a perspective based on the changing nature of science and that it is in turn open to criticism and further elaboration. Furthermore, there is no conflict between historical reconstructions and naturalistic approaches, as the historical approach furthers the naturalistic program.
2. No theory can claim to provide us with an objectively true picture of the world (Giere 2006a). All theoretical formulations are eventually superseded. Similarly, most purported laws of nature are false and do not help much in understanding scientific practice (Giere 1999).
3. The relationship between naturalism and cognitive science has been described by Giere (2010) in the following terms: "Examples of such cognitive processes include mental modeling, creating analogies, and devising thought experiments" (p. 217). Science educators have followed this line of research at least for the last three decades. One of the examples is provided by the information-processing theory of Pascual-Leone (1978), which has been applied extensively in science education (Lawson 1983; Niaz and Robinson 1992; Niaz 1988, 1989; Tsaparlis et al. 1998).
4. Heuristic principles constitute an important tool for facilitating domain-specific knowledge and are expressed in cogent terms by Schwab (1974): "In physics, we did not know from the beginning that the particles of matter are fundamental and determine the behavior of these particles; their relations to one another. It was not verified knowledge but a heuristic principle, needed to structure inquiry, that led us to investigate mass and charge and, later, spin" (p. 165). This recapitulates the evolution of the modern theory of atomic structure in the early twentieth century in succinct terms. Further examples are provided by Niaz (2012a, Chap. 2).
5. Conant's history of science, despite its Baconian overtones, can still be extended and invigorated by including episodes from the twentieth-century science and perhaps explore the possibility of microhistories.
6. Duschl and Grandy (2013) question the idea that "scientific knowledge is tentative" as this would undermine the confidence that scientific inquiry leads to scientific truths. These authors then refer to Newton-Smith's (1981) concept of "pessimistic induction" according to which any scientific theory once believed to

be true will eventually be found false. According to one reviewer of this book, the concept of pessimistic induction started with Laudan (1981). Next, these authors state, "We see this in Version 1 statements such as 'Scientific knowledge is tentative.' So, we might ask what confidence can we have in scientific inquiry leading to scientific truths?" (Duschl and Grandy 2013, p. 2130). To endorse their position, they quote Thagard (2007): "It is noteworthy that Laudan's examples are all from before the twentieth century, and one could argue that recent science has been more successful in achieving true theories" (p. 34). This statement can be questioned on two grounds: (a) Let us consider the following sequence, of atomic models all from twentieth century: J.J. Thomson → E. Rutherford → N. Bohr → Bohr–Sommerfeld → wave mechanical → recent advances. Are we to suggest to our students that some of these models were true and others false? (b) What exactly do we mean by "true theories"? Was Newton's theory true? Is it still true after Einstein's theory of relativity? And most important of all, is Einstein's theory here to stay and "true" forever? (cf. Giere 1999, 2006a). According to Worrall (2010): "On what grounds, then, could the realist deny the possibility that Einstein's theory might itself eventually be replaced by a theory bearing the same relation to it as it does to Newton's …" (p. 288, also see p. 281). It is concluded that the inclusion of tentative nature of science as part of the nature of science can help students to understand how science progresses and the difficulties involved.

7. The tentative nature of scientific knowledge is an extremely important issue for science education, as it would help our students think that science is an ongoing endeavor and in which they can also contribute by postulating new theories. Nevertheless, it is important to take note of what Hodson (2009) has referred to as "how tentative is tentative?":

   > It is important for students to recognize that scientific knowledge is tentative because it is based, ultimately, on empirical evidence that may be incomplete, and because it is collected and interpreted in terms of current theory, which may eventually be changed as a consequence of the very evidence that is collected. In all these endeavours, the creative imagination of individual scientists is impacted by all manner of personal experiences and values. Moreover, the 'collective wisdom' of the scientific community that supports the practice, scrutinizes the procedures and evaluates the products, is also subject to complex sociopolitical, economic and moral– ethical forces. In consequence, there can be no certainty about the knowledge produced (p. 164).

   After this statement, Hodson (2009) raises the issue of "how tentative is tentative" by emphasizing that we do not have to give students the impression that scientific theories are falsified by simple procedures of data collection and arguments. Precisely, this is the crux of the issue that involves the intervention of peer evaluation, collective wisdom of the scientific community, or what Campbell (1988) has referred to as *competitive cross-validation*.

8. Finally, based on the naturalistic reflection on the practice of science, Giere (2006a) espouses perspectivism: "… scientific knowledge claims are perspectival rather than absolutely objective. It follows almost immediately that some contingency is always present in any science. That human observation is per-

spectival, a function of an interaction between the world and human cognitive capacities, seems to me indisputable … Here there is need for what Kuhn (1962, Chap. 1) called 'a role for history.' Looking back historically, we can examine and understand the perspectival nature of earlier theoretical perspectives" (p. 93). Most science educators would find this statement quite novel and at the same time exemplifying our current scientific practice, based upon (a) perspectival rather than objective, (b) presence of contingency (for details, see Chap. 6), (c) role of human cognitive capacities, (d) role of history, and (e) recognition of earlier theoretical perspectives.

Chapter 3
Nature of Science in Science Education: An Integrated View

Introduction

What is science and how it progresses is an important part of understanding the nature of science (NOS). Philosophers of science have debated the subject at length, and similarly science educators have also contributed toward not only understanding the nature of NOS but rather how to introduce it in the classroom (Abd-El-Khalick 2005, 2012; Alters 1997; Blanco and Niaz 2014; Clough 2006; DiGiuseppe 2014; Eflin et al. 1999; Hodson and Wong 2014; Kang et al. 2005; Lederman et al. 2002; Matthews 2012; McComas and Olson 1998; Niaz 2001a; Osborne et al. 2003; Smith et al. 1997; Smith and Scharmann 1999, 2008; Vesterinen and Aksela 2013). According to Wolpert (1993), the nature of science is difficult to understand as it is rather unnatural:

> … both the ideas that science generates and the way in which science is carried out are entirely counter intuitive and against common sense—by which I mean that scientific ideas cannot be acquired by simple inspection of phenomena and that they are very often outside everyday experience. Science does not fit with our natural expectations. (p. 1)

Similarly, Matthews (2015) has emphasized the multidisciplinary nature of science and hence the difficulties involved in its introduction in the classroom:

> Science is a human, and thus *historically embedded*, truth-seeking enterprise that has many features: cognitive, social, commercial, cultural, political, structural, ethical, financial, psychological etc. All these features are worthy of study by science students, as well as by disciplinary specialists, and different ones come into clearer focus when considering different sciences, and when considering aspects of the history, achievements and practice of the different sciences. (p. 388, italics added)

In a series of studies, J.M. Campanario has drawn attention to the need for science students to understand the scientific endeavor. Most science curricula, courses, and textbooks in many parts of the world ignore that: (a) Many theories accepted today as integral part of scientific knowledge had to struggle against the skepticism

M. Niaz, *Chemistry Education and Contributions from History and Philosophy of Science*, Science: Philosophy, History and Education,
DOI 10.1007/978-3-319-26248-2_3

of the scientific community (Campanario 1993). (b) In some cases scientific papers later recognized and awarded the Nobel Prize originally had to face difficulties in order to be accepted for publication in the scientific journals (Campanario 1995). (c) Some of the most cited articles in the history of science were initially rejected by the reviewers of scientific journals (Campanario 1996). (d) Some of the most famous scientists resisted the acceptance of new theories as they seemed to question some of their own theories (Campanario 1998). These aspects contrast sharply with the stereotype of scientists who are often portrayed as working disinterestedly and above all looking for "truths." Based on his previous work (summarized above), Campanario (1999) concluded that there is *a science that we do not teach* to our students and thus deprive them from having an adequate vision of the nature of science.

In a study (Niaz 2012a, Chap. 4) designed to introduce the nature of science to in-service chemistry teachers, based on a history and philosophy of science perspective, the article by Campanario (1999) was used as a reading material. As part of their end-of-course evaluation, participants were asked the following question: "Do you think there is a science that we do not teach?" One of the teachers responded in the following terms:

> Of course there is a science that we do not teach, based on the premise that we teach science content and not "nature of science", (NOS), which makes understanding science more difficult. Issues related to NOS need to be debated in the classroom if we want our students to understand "science in the making". For example it would be very helpful and motivating for students if the following issues are discussed: Where does an original idea come from? Why do scientists have to publish their work? Is competition between scientists inevitable? Who finances research? And how to recognize the merit of a scientist? These issues can perhaps be included at the end of a chapter in sections such as: Novel aspects of science, New tendencies in research and education, Science for the future, etc. (Reproduced in Niaz 2012a, p. 142)

This, of course, raises many issues, such as: where does an original idea come from and whether competition and rivalry between scientists is inevitable. Furthermore, the relationship between "science content" and "nature of science" is the subject of considerable interest to science educators. Teaching "science content" (topics of the science curriculum found in textbooks) and leaving out "nature of science" (dynamics of scientific progress) inevitably leads to what Schwab (1962) has referred to as science being presented as a "rhetoric of conclusions." This leads to a dilemma, as emphasizing the rhetoric of conclusions precisely leads to memorization of science content. How do we go beyond that and look for new teaching strategies? According to Campanario (2002), just like scientists, students (and perhaps teachers) also offer resistance to new ideas, and the inclusion of controversial historical episodes (the metacognitive dimension of science) can stimulate students' intellectual curiosity. This, once again, leads to the dilemma mentioned in Chap. 2 that despite the naturalistic turn in the philosophy of science, do we still need reconstruction of historical episodes in the classroom or not? Chapter 2 provides a possible alternative in a section entitled "Conant and History of Science."

Nature of Science: Domain-General or Domain-Specific?

According to Duschl and Grandy (2013), the understanding of NOS can be considered as an ongoing debate between two groups of science educators, namely (see also Chap. 2):

> Version 1: "... the position that NOS should be benchmarked using domain-general, consensus-based aspects of NOS and taught through explicit references to a set of heuristic principles that philosophers of science and historians of science use to characterize science as a way of knowing (cf. Holton 1978a; Lakatos 1970; Laudan 1977)." (p. 2111)
>
> Version 2: "... the position that science, as well as science education, should be conceptualized in terms of cognitive, epistemic, and social practices (Giere 1988; Nersessian 2002) and the material and technological contexts (Pickering 1992) that characterize doing science. The Version 2 science education position is that NOS learning occurs when students' engagements are situated in these practices, in age appropriate contexts." (pp. 2111–2112)

After this classification, and by endorsing Version 2, Duschl and Grandy (2013) state:

> At the core of the debate is what comes to be seen as "explicit" teaching of NOS. Version 1 advocates that teachers explicitly link the consensus statements to features of science lessons and activities. Version 2 advocates students engage in domain-specific scientific practices during weeks or months long curriculum units that focus the learners' attention on the model building and refining enactments found in measuring, observing, arguing from evidence and explaining that are part of the growth of scientific knowledge. (p. 2112)

Duschl and Grandy do not provide further details with respect to what they mean by heuristic principles and how these relate to historical reconstructions. Within a history and philosophy of science perspective, heuristic principles represent a "deliberate construction of the mind" (Schwab 1974) that help to understand data, findings, and historical reconstructions of events. At this stage it is important to examine how Duschl and Grandy (2013) conceptualize the role played by heuristic principles and the underlying historical reconstructions in Version 1 of NOS: "... Version 1 embraces a rational reconstruction view of scientific developments and the growth of knowledge ... the consensus NOS statements ... become the teachable moments in the examination of physical science historical episodes (e.g., Bending of Light in the 1919 Eclipse Experiment: Einstein and Eddington; Kinetic theory: Maxwell's Presuppositions) [which] can be used for ... guiding science teaching about NOS" (p. 2122). After reading this, a science educator would like to know the teaching strategy followed by Version 2 when historical episodes are discussed in the classroom, namely, would they establish a relationship with NOS. If not, then what happens to the historical perspective of science (e.g., that of Conant), and if the answer is in the affirmative, then how does Version 1 differ from Version 2?

Let us now consider the assertion that Version 1 science educators follow the heuristic principles, which historians and philosopher of science (Holton, Lakatos, and Laudan) use to characterize science as a way of knowing. At this stage it would be interesting to consider the *oeuvre* of these three scholars and see how they elaborated their heuristic principles.

Holton's Heuristic Principles

In his *Scientific Imagination*, Holton (1978b, 1998) deals at length with various case studies based among others on the work of Newton, Einstein, Fermi, Millikan, and Ehrenhaft. Just one of these, the determination of the elementary electrical charge and the ensuing controversy between Millikan and Ehrenhaft, is considered to be a classic in historical reconstruction. In other words, heuristic principles derived by Holton were not elaborated a priori but were rather based on a domain-specific topic in the history of science, which interestingly is at the same time an important topic of the science curriculum in almost all parts of the world. In my opinion, some of the heuristic principles derived from this controversy are: (a) a reconstruction of events that led to the Millikan–Ehrenhaft controversy, (b) Millikan's guiding assumption (presuppositions), (c) suspension of disbelief, (d) transfer of charge as an integral multiple of the elementary electrical charge, (e) dependence of the elementary electrical charge on experimental variables, and (f) Millikan's experiments as part of a sequence of heuristic principles. These heuristic principles precisely constitute the criteria used by Niaz (2000a) to evaluate the presentation of the oil drop experiment in general chemistry textbooks published in the USA. A closer analysis of these criteria will show that the first three are fairly domain-general (applicable across domains), whereas the last three are almost entirely domain-specific. Furthermore, it is important to note that both the domain-general and domain-specific heuristic principles are not dichotomous but rather closely enmeshed with each other within a historical context.

Lakatos's Heuristic Principles

Lakatos's heuristic principles are based on the following domain-specific historical episodes: (a) Copernican astronomy, (b) Newtonian mechanics, (c) Prout's atomic theory, (d) Michelson–Morley experiment, (e) Bohr's model of the atom, (f) Bohr–Sommerfeld model of the atom, and (g) Einstein's theory of relativity. All essential aspects of Lakatosian methodology (research programs, negative/positive heuristics, protective belt, problem shifts, crucial experiments) are based on antecedents derived from a historical reconstruction of the domain-specific topics of the history of science. Furthermore, Lakatos provides a deeper understanding of the philosophy of science itself by using examples from the history of science. For example, according to Duhem (1914), the reasons for selecting a theory are based among other factors on "good sense." Lakatos goes beyond "good sense" by differentiating explicitly between the negative heuristic and the protective belt, and the latter can be modified in the light of empirical evidence, whereas the negative heuristic specifies what can be protected from refutation. This clearly shows that heuristic principles are not based on a priori normative principles but rather based on historical research itself (for a detailed analysis of the relationship between the philosophies of Duhem and Lakatos, see Niaz 2009, Chap. 3).

Laudan's Heuristic Principles

Kuhn (1962, 1970) has argued forcefully for abandoning our approach to history as a simple repository of anecdote and chronology, if we really want to understand the dynamics of scientific progress. Among others, Laudan et al. (1986, 1988) have followed this advice by studying a more varied set of case studies, such as Galileo, Newton, the chemical revolution, Kekulé (benzene theory), fermentation theory, polywater episode, Ampère's electrodynamics, Brownian motion, Michelson–Morley experiment, plate tectonics, theory of an expanding earth, and Planck's quantum crisis, among others. This is an ambitious agenda and has provided a gold mine of information with respect to how theories are constructed, developed, and changed. The chemical revolution as presented by Perrin (1988) provides greater insight with respect to what was already known. Kuhn (1962, 1970) had previously used the chemical revolution to suggest that a change in guiding assumptions is a process that is abrupt and total. Perrin's (1988) research, in contrast, reveals that Lavoisier's new guiding assumptions did not emerge full-blown but piece by piece over a decade or more.

Among other aspects, one that looms large in this whole study is that of "guiding assumptions" (similar to "presuppositions" for Holton, "hard core" for Lakatos, "paradigm" for Kuhn) and has considerable implications for the transition from historical to the naturalist stance. Different case studies have provided an agreement with respect to the following claims about guiding assumptions (Donovan et al. 1988): (a) Guiding assumptions are one of the most important units for understanding scientific change. (b) Guiding assumptions, once accepted, are rarely abandoned just on the basis of negative empirical evidence. This is at odds with the Popperian insistence on the centrality of refutation. Even working scientists ignore that counterevidence does not necessarily introduce changes in a theory and thus provide instant rationality. (c) Observations are theory-laden. (d) Guiding assumptions are not abandoned unless there is a new candidate to replace them. (e) Coexistence of rival sets of guiding assumptions is the rule rather than the exception. (f) There is a constant debate and controversy among rival sets of guiding assumptions. (g) Metaphysical and other nonscientific factors play an important role in the assessment of scientific theories and guiding assumptions.

It may be no surprise for someone following historians and philosophers of science that the thesis of Laudan et al. (1986, 1988) and Donovan et al. (1988) has been questioned and is a subject of controversy. Howson (1990) considers this thesis to be nothing less than a "manifesto" of the "historical school" and questions its methodology and some of the conclusions. The main argument is that by looking for and accepting empirical evaluation through case studies amounts to a positivistic framework. Thus, negative empirical evidence would lead to rejection of the theses of the historical school. Without going into the merits of Howson's arguments, it is important to note that if science itself progresses through arguments, counterarguments, and controversies, no wonder philosophy of science has also chosen the same path. To make matters more difficult, Howson (1976), in his early work, had also contributed to the development of the historical school.

Interestingly, according to Giere (2010, p. 223), Laudan's (1996) *Beyond Positivism and Relativism* can also be considered as *normative naturalism.* Let us see how Laudan conceives the role of history (contrast this with Howson) in his normative naturalism:

> *Because our aims and background beliefs differ from those of past scientists, determinations of the rationality of their actions and of the soundness of our methodological proposals cannot be collapsed into one and the same process. Rationality is one thing: methodological soundness is quite another.* Since that is so, the historicist's rejection of the methodological enterprise, like his rejection of specific methodologies, on the grounds that they render the history of science irrational is a massive non sequitur. (Laudan 1996, p. 131, italics in the original)

This clearly shows that Laudan accepts naturalism and at the same still recognizes the role of history, quite similar to the views of both Phillips (2014) and Siegel (2014) discussed in Chap. 2. Laudan then goes on to explain that although the historicists generally expect that in most cases science has been a "quintessentially rational activity" still in the case of a conflict between the methodology and the actual decisions made by the Newtons, Einsteins, and Darwins of the world, it is the methodology that has failed and not that the scientists were irrational. As a master stroke, Laudan (1996) concluded:

> Virtually all the philosophers in this camp would subscribe to the spirit of Lakatos's assertion that: "A rationality theory [by which Lakatos specifically means a methodology] … is to be rejected if it is inconsistent with an accepted 'basic value judgement' of the scientific elite." (p. 127, material within brackets by Laudan; underlined material reproduced from Lakatos 1971, p. 110)

This helps us to understand the historical reconstructions vis-à-vis the naturalist turn in the philosophy of science. In other words, the historical reconstruction (methodology) is itself naturalistic (Siegel 2014), and Lakatos and others can even be classified as naturalistic, subject of course to critical appraisals (Phillips 2014). Interestingly, Lakatos (1971) himself recognized the need for improving reconstructions:

> … no such historiographical research programme can or should explain *all* history of science as rational: even the greatest scientists make false steps and fail in their judgment. Because of this *rational reconstructions remain for ever submerged in an ocean of anomalies. These anomalies will eventually have to be explained either by some better rational reconstructions or by some "external" empirical theory.* (p. 118, italics in the original)

Domain-General or Domain-Specific Nature of Science: A False Dichotomy

In the previous section, I showed that the domain-general heuristic principles formulated by philosophers of science (Holton, Lakatos, and Laudan) are themselves derived from an in-depth (domain-specific) historical reconstructions of particular episodes in the history of science. One reviewer of this book disagreed and suggested that Laudan and especially Lakatos imposed their methodology on their case studies. Interestingly, Lakatos (1971, p. 118) even recognized that historical reconstructions are themselves open to revision. (See the previous section with respect to how Lakatos and Laudan elaborated their heuristic principles based on a historical reconstruction of various episodes in the history of science). For further details with respect to Lakatos's methodology, see Niaz (2009), especially Chap. 3 entitled "Understanding scientific progress: From Duhem to Lakatos." In this section, I provide examples from research in science education to show that both the domain-general and domain-specific aspects of nature of science (found in some disciplines only) complement each other and are essential for facilitating students' conceptual understanding, based on an integration of the two (see Table 3.1).

Empirical Nature of Science

Thomson (1897) is generally credited to have "discovered" the electron while doing experiments with cathode rays. Determination of the mass-to-charge (*m*/e) ratio of the cathode rays can be considered as the most important experimental contribution of Thomson. Yet, he was neither the first to do so nor the only experimental physicist to work on this. Kaufmann (1897) and Wiechert (1897) also determined the *m/e* ratio of cathode rays in the same year as Thomson, and their values agreed with each other. If we tell students that "science is empirical," we shall be denying students an important aspect of the nature of science, namely, what made Thomson's work different from that of Kaufmann and Wiechert (Niaz 1998b). Falconer (1987) has provided insight on the subject: "Kaufmann, an ether theorist, was unable to make anything of his results. Wiechert, while realizing that cathode ray particles were extremely small and universal, lacked Thomson's tendency to speculation. He could not make the bold, unsubstantiated leap, to the idea that particles were constituents of atoms" (p. 251). The rationale behind the empirical determination of the *m/e* ratio was provided by the heuristic principle, namely, Thomson decided to determine the ratio to identify cathode rays as ions (if the ratio was not constant) or as universal charged particles (if the ratio was constant). In other words, Thomson did the experiment based on a deliberate conception of the mind.

Similarly, in the case of Robert Millikan's oil drop experiment, it was the electron theory which suggested the existence of the elementary electrical charge and hence the need for its experimental determination (for details, see Holton 1978a;

Table 3.1 Integration of domain-general and domain-specific aspects in the history of science (a summary)

Domain-general aspects of NOS	Domain-specific aspects of NOS	Reference
Empirical nature of science	Determination of mass-to-charge (m/e) ratio of cathode rays	Falconer (1987)
	Oil drop experiment	Holton (1978a)
Objectivity in science	Award of Nobel Prize to Millikan	Holton (1988)
	Thomson's prestige and controversy with Rutherford	Wilson (1983)
Competition among rival theories	Thomson and Rutherford models of the atom	Wilson (1983) Heilbron (1981)
	Lenard, Planck, and the Einstein hypotheses to explain photoelectric effect	Wheaton (1983)
	Copenhagen, Schrödinger, and de Broglie hypotheses of quantum mechanics	Cushing (1998)
		Tsaparlis (2001)
	Valence bond and molecular orbital models for explaining chemical bonding	Gavroglu and Simões (2012)
		Giere (2006a, b) Hoffmann et al. (2003)
Different interpretations of the same experimental data leading to controversies	Thomson and Rutherford interpretations of alpha-particle experiments	Heilbron (1981) Wilson (1983)
	Millikan and Ehrenhaft interpretations of the oil drop experiment	Holton (1978a)
	Arrhenius, Brönsted–Lowry, and Lewis interpretations of acid–base reactions	Furió-Más et al. (2005)
	Concept of osmosis based on thermodynamics and kinetic–molecular approach	De Berg (2006)
	Statistical and phenomenological models to understand thermodynamics	Tarsitini and Vicentini (1996)
	Electrolyte solution chemistry: explanation of osmotic pressure by the ionist and the hydrationist schools	De Berg (2014)
		Heyrovska (1996)
Inconsistent nature of scientific theories	Maxwell's kinetic theory	Brush (1976)
	Bohr's model of the atom	Cooper (1992)
		Vickers (2013)
Role of refutation or falsification	Millikan's determination of Planck's constant *h* and belief in wave theory of light	Holton (1999)
		Lakatos (1970)
	Michelson–Morley experiment	Lakatos (1970) and Holton (1969)
	Synthesis of urea by Wöhler	Ramberg (2000)

(continued)

Table 3.1 (continued)

Domain-general aspects of NOS	Domain-specific aspects of NOS	Reference
Theory-laden nature of observations	Millikan's presupposition: atomic nature of electricity. Ehrenhaft's presupposition: anti-atomism	Holton (1978a)
	Dyson and Eddington's presupposition: Einstein's theory of general relativity	Collins and Pinch (1993)
		Earman and Glymour (1980) Niaz (2009)
Tentative nature of scientific knowledge	Atomic models in the twentieth century	Wilson (1983) Niaz and Cardellini (2011)
	Lewis structures, valence bond, and molecular orbital models of chemical bonding	Lewis (1923) Pauling (1931) Mulliken (1932)
	Arrhenius, Brönsted–Lowry, and Lewis models for understanding acid–base phenomena	De Berg (2003)
		Kousathana et al. (2005)
	From Newtonian mechanics to Einstein's theory of relativity	Giere (2006a)
		Worrall (2010)
Scientific ideas are affected by their social and historic milieu (*milieu of the time*)	Michelson–Morley experiment	Cooper (1992)
		Weingart (2004)
		Longino (1990), Machamer and Wolters (2004)
Systematicity	Demarcation of real science from pseudoscience	Hoyningen-Huene (2013)
		Ruse (2014)

Niaz 2000a, 2009). Thus, the experiment and the empirical data became important in the light of a heuristic principle, namely, the corroboration of the atomic nature of electricity. In other words, this principle helped Millikan to understand the data. According to Van Fraassen (1980), Millikan was writing theory with his experimental apparatus. On the contrary, Achinstein (1991) has argued that it was the experiment that led Millikan to postulate the elementary electrical charge.

Objectivity in Science

History of science shows that the plausibility of some conceptions (theory/generalization) can be upheld even if there is evidence to the contrary. In other words, a community of researchers may decide to interpret empirical evidence in a particular way, even if there are alternative interpretations available. Campbell (1988), an epistemologist, has expressed this concern in the context of the problematic nature of objectivity in science:

> The objectivity of physical science does not come from turning over the running of experiments to people who could not care less about the outcome, nor from having a separate staff to read the meters. It comes from a social process that can be called *competitive cross-validation* … and from the fact that there are many independent decision makers capable of rerunning an experiment, at least in a theoretically essential form. (p. 324, italics added)

The role of "many independent decision makers" is precisely that of the scientific community. For example, in the oil drop experiment, despite the availability of experimental evidence (starting in 1910), it took many years for the scientific community to reach consensus, which finally led to the award of the Nobel Prize to Millikan in 1923. According to one reviewer of this book, "It's worth pointing out though that there were no 'independent decision makers' who rerun those experiments (oil drop). To put it another way, there was no proof that Ehrenhaft's experiments were flawed." This is a complex issue and merits further discussion. Within a historical perspective, Daston and Galison (2007) have explored the complexities of the issues involved in the following terms:

> To grant objectivity a history is also to historicize the framework within which much philosophy, sociology, and history of science has been cast in recent decades. The opposition between science as a set of rules and algorithms rigidly followed versus science as tacit knowledge (Michael Polanyi with a heavy dose of the later Ludwig Wittgenstein) no longer looks like the confrontation between an official ideology of scientists as supported by logical positivist philosophers versus the facts about how science is actually done as discovered by sociologists and historians. Instead, both sides of the opposition emerge as ideals and practices with their own histories—what we have called mechanical objectivity and trained judgment." (p. 377)

Indeed, this sets the stage for understanding progress in science within a much richer context, in which mechanical objectivity would approximate to the ideals of logical positivism and objectivity as trained judgment to how science is actually done. Interestingly, Daston and Galison (2007, p. 478) consider the resolution of the controversy with respect to the determination of the elementary electrical charge between Millikan and Ehrenhaft (Holton 1978a), as an example of trained judgment.

Despite J.J. Thomson's prestige, the controversy with E. Rutherford with respect to the alpha-particle scattering experiments lasted for many years before Rutherford's single-scattering hypothesis was accepted by the scientific community.

The interpretation of observational data based on the bending of light in 1919 eclipse experiments was extremely difficult (Dyson et al. 1920). In order to understand the dilemma, Niaz (2009) has suggested a hypothetical scenario:

> [Let us suppose that] Eddington and Dyson are not aware of Einstein's General Theory of Relativity and particularly of the prediction that sunlight near the sun would bend. Under these circumstances experimental evidence from all three sources (Sobral and Principe) would have been extremely uncertain, equivocal, and difficult to interpret. (p. 135)

Results of these observational data were presented at a joint meeting of the Royal Society and the Royal Astronomical Society on November 6, 1919. The meeting was chaired by J.J. Thomson, President of the Royal Society, who "endorsed" the results in the following terms:

> *It is difficult for the audience to weigh fully the meaning of the figures* that have been put before us, but the Astronomer Royal [Dyson] and Prof. Eddington have studied the material carefully, and they regard the evidence as decisively in favour of the larger value for the displacement. This is the most important result obtained in connection with the theory of gravitation since Newton's day." (Thomson 1919, italics added)

Based on these considerations, it is plausible to suggest that: (a) Science is objective only in a certain context (perspectival, according to Giere 2006a) of scientific development. (b) Objectivity in science is based on a social process of competitive cross-validation through critical peer evaluation. (c) Science is not characterized by its objectivity but rather by its progressive nature (Lakatos 1970; Laudan 1996). For details with respect to the changing nature of objectivity in the history of science, see Daston and Galison (2007).

Competition Among Rival Theories

Rutherford's (1911) experiments with alpha particles and the resulting model of the nuclear atom had to compete with a rival framework, namely, Thomson's model of the atom, referred to as "plum-pudding" in most textbooks. The rivalry between the Thomson and the Rutherford models of the atom was well known in scientific circles, and Rutherford in a letter dated Feb. 2, 1914, to the Secretary of the Royal Society had this to say: "I have promulgated views on which J.J. [Thomson] is, or pretends to be, skeptical. At the same time I think that if he had not put forward a theoretical atom himself, he would have come round long ago, for the evidence is very strongly against him" (reproduced in Wilson 1983, p. 338).

In the first decades of the twentieth century, P. Lenard's trigger hypothesis, M. Planck's quantum analog of the trigger hypothesis, and A. Einstein's quantum hypothesis constituted rival hypotheses to explain the same photoelectric phenomenon (for details see Niaz et al. 2010a). This led to considerable controversy, and even Planck, the founder of the quantum hypothesis, opposed Einstein.

Around 1927, besides the Copenhagen interpretation of quantum mechanics, there were two rival interpretations, namely, Schrödinger's wave picture and de Broglie's pilot-wave model—a precursor to Bohm's theory of hidden variables (for details see Cushing 1998 and Chap. 6). According to Tsaparlis (2001):

> The Schrödinger equation is the foundation of quantum mechanics. All the effort in quantum physics and quantum chemistry research lies in formulating approximate solutions to this equation, since exact solutions do not exist except for some simple systems. Therefore, various approximations are in fashion one time or another, and those which are good for today may not be so for tomorrow. One thing however is certain: the Schrödinger equation will still be the starting point for any new improvement. (p. 211)

This clearly shows not only the rivalry between different approaches to quantum mechanics but also the need for a continual critical appraisals leading to new formulations.

Starting in the 1930s, chemists have been developing the valence bond and molecular orbital models for understanding and explaining chemical bonding. Both models are based on quantum mechanics and at best can be considered as approximations. The rivalry between the two models facilitates our understanding of progress in science as perspectival rather than objective (for details see Giere 2006a, Gavroglu and Simões 2012, Hoffmann et al. 2003, also Chap. 6).

Different Interpretations of the Same Experimental Data Leading to Controversies

Both J.J. Thomson and E. Rutherford had very similar experimental results based on the scattering of alpha particles. Based on these empirical findings, Thomson propounded the hypothesis of compound scattering (multitude of small scatterings), whereas Rutherford propounded the hypothesis of single scattering, in order to explain the large-angle deviations of alpha particles. The two interpretations led to two different models of the atom and a bitter dispute between the proponents (for details see Heilbron 1981, Niaz 1998b, and also Chap. 2).

In the determination of the elementary electrical charge, R.A. Millikan and F. Ehrenhaft obtained very similar experimental results, and yet their guiding assumptions led them to postulate the electron and sub-electron (fractional charge), respectively, and the controversy lasted for many years (1910–1923). According to Holton (1978a), "It appeared that the same observational record could be used to demonstrate the plausibility of two diametrically opposite theories, held with great conviction by two well-equipped proponents and their respective collaborators …" (pp. 199–200). For details see Niaz (2000a) and also Chap. 2.

Historically, one of the first models to understand the acid–base reactions was based on empirical knowledge of electrical conductivity of ionic solutions. Although Arrhenius first presented his model in 1887 based on the ionic dissociation hypothesis, it was not well received. It was only in the twentieth century that three models (Arrhenius, Brönsted–Lowry, and Lewis) were fully postulated consecutively and even today are presented and utilized in most high school and university general chemistry textbooks simultaneously to understand different aspects of the acid–base phenomena (Furió-Más et al. 2005). Developments of acid–base models also manifest another nature of science facet, namely, the tentative nature of scientific knowledge (see a later section in this chapter).

In the middle of the twentieth century, mathematical models based on thermodynamics were postulated to understand the phenomenon of osmosis (J.H. van't Hoff and others). Later by the end of the twentieth century, the kinetic–molecular approach to osmosis was developed (by U. Lachish and others). At present both models are presented and used in general chemistry textbooks (for details see De Berg 2006).

Currently, most general chemistry and physics textbooks present two rival approaches to understand thermodynamics: (a) a statistical approach based on the ideas of J.C. Maxwell and others, which is based on abstract models, and (b) a phenomenological approach based on the ideas of M. Planck and others, which is more directly related to experience (Tarsitani and Vicentini 1996).

The development of electrolyte solution chemistry, starting in the late nineteenth and early twentieth century, was the subject of considerable controversy (De Berg 2014b). The ionist school (S. Arrhenius, W. Ostwald, J. van't Hoff) postulated that the measurement of osmotic pressure of a range of aqueous salt solutions was best interpreted in terms of the partial dissociation of the dissolved substance. van't Hoff was the recipient of the first Nobel Prize in chemistry in 1901. On the other hand, the hydrationist school (H. Armstrong and S. Pickering) maintained that the production of hydrated compounds from solution was evidence that the solution was an *association with water* rather than *dissociation in water* phenomenon. Interestingly, recent developments in electrolyte solution chemistry have drawn on the interpretations of both the ionists and the hydrationists (Heyrovska 1996), thus providing further support to the thesis that the same experimental data can be interpreted differently. De Berg (2014) has argued cogently that this historical episode can provide students with an insight with respect to the nature of science provided that both textbooks and teachers explicitly use the historical context:

> This is a little like teaching the solution process as a combination of hydration and dissociation without realizing or identifying that there is a rich historical context of controversy behind the phenomenon of solution, and it is this rich historical context that tells us something of what the practice of science is like. (p. 5)

Again, this illustrates the dilemma faced by science educators: are we to simply teach science content without any recourse to the "practice of science"?

Inconsistent Nature of Scientific Theories

Maxwell's (1860) seminal paper on the kinetic theory is a good example of a research program progressing on inconsistent foundations. On the one hand, it was based on "strict mechanical principles" derived from Newtonian mechanics, and still at least two of Maxwell's simplifying assumptions (referring to the movement of particles and the consequent generation of pressure) were in contradiction with Newton's hypothesis explaining the gas laws based on repulsive forces between particles. Brush (1976) has pointed out the contradiction explicitly: "… Newton's laws of mechanics were ultimately the basis of the kinetic theory of gases, though this theory had to compete with the repulsive theory attributed to Newton" (p. 14). For details see Niaz (2000b).

Bohr's (1913) model of the atom incorporated Planck's "quantum of action" to the classical electrodynamics of Maxwell. For many of Bohr's contemporaries and philosophers of science, this represented a contradictory "graft" or an inconsistent foundation. For further details see Niaz (1998b) and also Chap. 2 (especially the views of the following three scholars: Cooper 1992; Darrigol 2009; Margenau 1950).

The Role of Refutation and Falsification

The role of refutations and falsifications has been the subject of considerable debate in the history and philosophy of science literature. According to Lakatos (1970), a scientist generally does not abandon his "hard core" of beliefs or presuppositions in the face of anomalous data. After working on the photoelectric effect for many years, Millikan provided an experimental value of Planck's constant *h* based on Einstein's equation. Millikan, however, strongly believed in the wave theory of light and thus rejected Einstein's hypothesis of light quanta for many years (Holton 1999). This clearly shows that empirical data do not necessarily provide instant rationality by refuting a scientific theory.

The Michelson–Morley experiment is another good example to understand the role of refutations in the history of science. This experiment, first conducted in 1887, provided a "null" result with respect to the ether-drift hypothesis, namely, that there was no observable velocity of the earth with respect to the ether. Despite considerable experimental evidence, it took almost 25 years for this hypothesis to be refuted and recognized as the "greatest negative experiment in the history of science" (Lakatos 1970, p. 162). Textbooks published in different countries even today generally emphasize that it was the Michelson–Morley experiment that led Einstein to postulate his special theory of relativity (Arriasssecq and Greca 2007; Brush 2000).

Synthesis of urea in 1828 by F. Wöhler (1800–1882) constitutes an important episode in the history of organic chemistry (Ramberg 2000). It was expected that the reaction between cyanic acid and ammonia should produce ammonium cyanate, a salt. However, Wöhler found that urea was not a salt and it did not have any of the properties of cyanates. Some of Wöhler's contemporaries considered the synthesis to be "epochal" and that he had synthesized urea from the elements and that this unified organic and inorganic chemistry under the same laws. Furthermore, it was believed that the synthesis of urea destroyed or at least weakened the theory of "vital force" in living organisms. According to vitalism organic matter possessed a special force or vital force inherent to all things living. Although Wöhler himself was reluctant to accept that his synthesis had falsified the theory of vital force, it still constitutes a myth not only in organic chemistry textbooks but also for some historians of chemistry (cf. Ramberg 2015).

Most philosophers of science would agree that the relationship between theory and experiment is complex and a single experiment cannot falsify a theory. Ramberg

(2000) surveyed 33 organic chemistry textbooks published in the USA (between 1922 and 1926) and found that 91 % espoused the aforementioned myth by accepting some form of falsification. More recently, one general chemistry textbook even attributed Wöhler's success to the scientific method: "His [Wöhler's] experiment, however, produced an unexpected substance, which out of curiosity he analyzed and found to be urea (a constituent of urine). This was an exciting discovery, because it was the first time anyone had knowingly ever made a substance produced only by living creatures from a chemical not having a life origin. The fact that this could be done led to the beginning of a whole branch of chemistry called *organic chemistry*. Yet, had it not been for Wöhler's curiosity and his application of the scientific method to his unexpected results, the significance of his experiment might have gone unnoticed" (Brady et al. 2000, p. 3, original italics, underline added).

Theory-Laden Nature of Observations

Most historians and philosophers of science would agree with some form of the thesis that experimental observations are theory-laden. In other words, before doing an experiment, a scientist looks for its rationale, and this inevitably leads to the formulation of presuppositions/guiding assumptions (although these may not be explicitly formulated). According to Leon Cooper, experiments are difficult to perform and their meaning is elusive. This dilemma paves the way for a scientist to integrate data, theory, and conjecture, and in doing so, the role played by presuppositions is legitimized (cf. Niaz et al. 2010b).

One of the best examples of this thesis is provided by R. Millikan's presuppositions with respect to the atomic nature of electricity, which led to considerable controversy with F. Ehrenhaft. Interestingly, it is generally ignored that Ehrenhaft also had his presuppositions as he was sympathetic to the anti-atomistic ideas of E. Mach and thus interpreted his data accordingly (for details see Holton 1978a; Niaz 2005).

Another example is provided by F. Dyson and A. Eddington's interpretation of observations based on the deflection of light by the sun's gravitational field in the 1919 eclipse expedition. If it were not for Einstein's general theory of relativity as a presupposition/guiding assumption, it would have been extremely difficult to make sense of the observational data (for details see Niaz 2009, Chap. 9). Despite difficulties with the observational data, Dyson and Eddington were fully aware as to where the theory (Einstein's) was leading them. Following any other alternative course involved the risk of having all their observational data being considered ambiguous and ultimately invalid. This weighed more in this case as eclipse observations are not easy to replicate (contrast this with Millikan's oil drop experiment).

The theory-laden nature of observations was nicely summarized by C. Darwin (1861), who stated a complex idea in very simple terms: "How odd it is that anyone should not see that all observations must be for or against some view if it is to be of any service" (letter to Henry Fawcett, September 18, 1861, in Charles Darwin, Collected correspondence, 21 volumes. Cambridge University Press, Vol 9, p. 269).

Tentative Nature of Scientific Knowledge

The history of science shows that a continual critical appraisal of the scientific endeavor generally leads to theories/models with greater explanatory power. In the case of atomic structure, even in the twentieth century, scientists have developed a series of models that continue to provide increasing explanatory power, such as Thomson's, Rutherford's, Bohr's, Bohr–Sommerfeld's, and wave-mechanical, among others (for details see Chap. 4).

Chemical bonding has been the subject of considerable research and postulation of various models. In 1916, G.N. Lewis, based on the cubic atom as a theoretical device, explained the sharing of electrons to form the covalent bond. This led to the distribution of electrons in very simple molecules. After 1925, Pauli's exclusion principle provided a better explanation of how two electrons having the same charge, but different spin, can form a covalent bond. Based on quantum mechanics in the 1930s, Pauling and Mulliken presented the valence bond and the molecular orbital models, respectively. In 1957, Gillespie and Nyholm postulated the valence-shell electron-pair repulsion (VSEPR) model, which is generally considered to be a part of the valence bond model. This clearly shows how scientific knowledge with respect to chemical bonding has evolved and thus is considered to be tentative (for details see Chap. 6).

Different models for understanding acid–base reactions can also be considered as a manifestation of the tentative nature of scientific knowledge (see previous section on different interpretations of the same experimental data). The Arrhenius model is based on an ionic interpretation of electrical conductivity (De Berg 2003; Kousathana et al. 2005), and results from freezing point depression, osmotic pressure, and vapor pressure lowering experiments support it. The Brönsted–Lowry model proposed a broader definition of acids and bases, independent of their behavior in water, as the acidic and basic properties are considered to be independent of the solvent. Lewis amplified the model of acids and bases by introducing the concept of electron-pair donor (base) and acceptor (acid). This shows that the same topic may manifest two different facets of the nature of science.

The most famous example is of course the transition from Newtonian mechanics to Einstein's theory of relativity, which is the subject of a study designed to introduce the nature of science to in-service teachers, included later in this chapter. Furthermore, can we assure our students that Einstein's theory is the final word on this subject (for details see Chap. 2 and Worrall 2010)? This can be the subject of an interesting discussion in the classroom.

Scientific Ideas Are Affected by Their Social and Historic Milieu

Scientific knowledge is socially negotiated and needs to be understood in the historical milieu in which the experiments were carried out. This need not be confused with relativistic notions of science. According to Leon Cooper, it is the *milieu of the time* that helps us to understand why certain questions are asked and certain experiments are conducted. For example, as suggested by Cooper, if the Michelson–Morley (M–M) experiment had been done at the time of Copernicus, its result would have had no significance for the astronomers, as they considered the earth to stand still and at the center of the universe. (Of course, in order to make his point more explicit, Cooper is using a hypothetical example. The M–M experiment could not have been performed at the time of Copernicus, because the relevant technological and theoretical knowledge was missing.) Consequently: "It seems obvious that questions take their meaning in the context of what people believe at the time, and if you don't communicate this, you really are not communicating why people did things, why it was difficult, and how the ideas that we now accept evolved" (Cooper 1992, reproduced in Niaz et al. 2010b, p. 45). Understandably, the history of science, if and when included in the textbooks and classroom, needs to be interpreted within the context of the *milieu of the time* and not in hindsight.

Weingart (2004), a philosopher of science, has argued cogently about how scientific knowledge originates and gets validated through a social process until it forms part of the corpus of science:

> What starts out as the discovery of an individual or a small group of researchers may or may not become certified knowledge of the entire community. That process involves time, and, above all, the time required to criticize, test, and accept or reject the initial truth claim. In the interim the status of that claim is tenuous; it is subject to judgments that are heavily influenced by social rather than intellectual criteria: status of authority, prestige of institution, proximity of field, personal acquaintance, and so on. This is most obvious and well researched in the case of "peer review," that is, the very process of certifying knowledge by open criticism …. (pp. 114–115)

Other philosophers of science have also emphasized the interaction between science, values, and objectivity (Longino 1990; Machamer and Wolters 2004). Similarly, in science education the double-blind peer-review process is also considered as a social process required for validating research (Abd-El-Khalick et al. 2008).

Systematicity

Although the idea of systematicity in science has been used in the past by historians and philosophers of science, Paul Hoyningen-Huene (2013) has used it to characterize nature of science: "*Scientific knowledge differs from other kinds of knowledge, in particular from everyday knowledge, primarily by being more systematic*" (p. 14, italics in the original). Thus, the defining characteristic that differentiates science from other forms of knowledge is its superior systematicity, which in turn depends on the following nine dimensions: description, explanation, prediction, the defense of knowledge claims, critical discourse, epistemic connectedness, an ideal of completeness, knowledge generation, and the representation of knowledge (p. 27). Despite the importance of the systematicity thesis, philosopher of science Michael Ruse (2014) is somewhat skeptical: "… it was these people like Kuhn and Popper who pointed us away from the kind of philosophy of science that Hoyningen-Huene practices, namely, philosophy of science that is basically theoretical and divorced from the real science of the day …" (p. 285). Interestingly, however, the book has been more favorably received in the science education literature: "Hoyningen-Huene's comprehensive explication of the notion of systematicity offers a broad and dynamic picture of the *heterogeneous nature of science* and suggests that science is to be viewed as a family-resemblance concept. This book thus may offer an impulse for the discussion on consensus views that emphasize certain characteristics of science, such as tentativeness, for educational purposes (Lederman et al. 2002; Osborne et al. 2003) and alternative views that emphasize the heterogeneity of science and invoke the notion of science as a family-resemblance concept (Irzik and Nola 2011; van Dijk 2011)" (van Dijk 2013, p. 2372, italics added). Indeed, this contrasts with the views of Ruse (2014), who considers systematicity not to be based on the real episodes in the history of science, and hence would reflect more of a domain-general approach toward the understanding of the nature of science.

This picture becomes more blurred if we consider the following statement from Hoyningen-Huene (2013): "… the fact remains that there is no consensus among philosophers or historians or scientists about the nature of science at the beginning of the twenty-first century" (p. 5). If there is no consensus with respect to the nature of science (blurring of the picture), it means that the same is true of the *heterogeneous nature of science*. In other words, both domain-general and domain-specific (actual episodes in the history of science that are ignored by Hoyningen-Huene) aspects of the nature of science are important. Furthermore, it is precisely a historical reconstruction of the real science that shows the tentative nature of science (e.g., the changing nature of atomic models). Consequently, in science education we have three options: (a) emphasize domain-general aspects of the nature of science; (b) emphasize domain-specific aspects of the nature of science; and (c) integrate both aspects of the nature of science, which can help to integrate, for example, systematicity with real episodes in the history of science that are relevant for the science curriculum. From the perspective of the science curriculum and classroom practice, all aspects of nature of science are important; however, it is the domain-specific content (actual history) that remains paramount for understanding science and its progress. This issue is discussed further in Chap. 8.

Different Views of Nature of Science: Consensus, Family Resemblance, and Integrated View

At present there is considerable controversy among science educators with respect to understanding the nature of science (NOS). In order to facilitate understating, in this section, I present three different views of NOS: the consensus view, family resemblance view, and the integrated view.

The Consensus View

Based on a critical review of the science standards documents and the relevant history and philosophy of science literature, this view fosters a consensus among different research communities. It attempts to include in the classroom only those domain-general NOS aspects that are the least controversial. According to Lederman (2004):

> ... no consensus presently exists among philosophers of science, historians of science, scientists, and science educators on a specific definition of NOS. Hence, the reason for not placing the word "the" in front of NOS [Interestingly, an Editor at a leading international publisher told me that not to put "the" is incorrect]. This lack of consensus, however, should neither be disconcerting nor surprising given the multifaceted nature and complexity of the scientific endeavor. Conceptions of NOS have changed throughout the development of science and systematic thinking about science and are reflected in the ways the scientific education communities have defined the phrase "nature of science" during the past 100 years. (p. 303)

This view enjoys considerable support in the science education community and is perhaps a reasonable strategy in the face of considerable disagreements among historians, philosophers of science, and even science educators (Cobern and Loving 2001; Flick and Lederman 2004; Lederman et al. 2002; McComas et al. 1998; Osborne et al., 2003; Smith and Scharmann 1999; Zeidler et al. 2002).

Given the multidisciplinary characteristics of science, according to Matthews (2015), "... it is useful to understand NOS, not as some list of necessary and sufficient conditions for a practice to be scientific, but rather as something that, following Wittgenstein's terminology, identifies a 'family resemblance' of features that warrant different enterprises being called scientific" (p. 388). "List of necessary and sufficient conditions" refers to the "Lederman seven" list of NOS aspects (cf. Lederman et al. 2002). After introducing the idea of "family resemblance" Matthews (2015) recommends a change of terminology and research focus based on contextual and heterogeneous "features of science" (FOS), which are quite similar to the *consensus view*, except for the "... assumption that NOS learning can be judged and assessed by students' capacity to identify some number of declarative statement about NOS" (p. 389). This criticism pertains to the NOS construct and also the assessment of NOS, which varies considerably from one context to another, as one instrument may be appropriate for one classroom but not for another.

Family Resemblance View

Irzik and Nola (2011) have presented the family resemblance view of the nature of science based primarily on cognitive aspects: "It enables the teacher to characterize science in a nutshell as follows: *science is a cognitive system or pattern of practice and thought that involves such and such activities; values and aims at such and such; produces so and so using such and such methodologies and methodological rules*. Needless to say, such a characterization is only as good as the use it is put to, and we hope that we provided enough content and detail for it to be informative and enlightening for the purposes of science education" (p. 605, italics in the original). In order to implement this view in the classroom, the authors recommend the following approach: "For instance, the teacher may begin by asking what scientists do. This question is likely to prompt the students to come up with examples that fall under the category 'scientific activity.' Suppose observing, experimenting and theory building came up. A host of interesting questions can be pursued in this context: Is observing a passive activity? How does observation differ from experimentation? What is the point of doing an experiment and how does it relate to theory?" (Irzik and Nola 2011, p. 603). It is precisely such questions that led Matthews (2012, p. 4) to critique the consensus view of the nature of science, namely, the assumption that NOS learning can be judged and assessed by students' capacity to identify some number of declarative statements about NOS. Furthermore, a science teacher may be struck by the almost complete absence in the family resemblance view of the actual science content, namely, the history of science that forms such an important part of the science curriculum in almost all parts of the world. For example: (a) What was J.J. Thomson looking for when he designed his cathode ray experiments? (b) How did E. Rutherford's experimental findings changed his view of the atomic model? (c) Both Thomson and Rutherford had very similar experimental data and still their interpretations were entirely different. (d) How and why was Millikan led to the oil drop experiment? In my opinion, in contrast to the questions posed by Irzik and Nola (above), it is these questions that may arouse students' interest and hence facilitate the introduction of the nature of science. Interestingly, the authors themselves recognize this aspect of their view and justify it in the following terms: "The attentive reader will notice that we have said very little about the social embeddedness of science. This is because our paper has focused on the cognitive aspects of science" (p. 605).

Integrated View

A preliminary version of the integrated view was first presented by Niaz (2001a): "It is concluded that nature of science manifests in the different topics [domain-specific] of the science curriculum as heuristic principles. Science education, by emphasizing not only the empirical nature of science [domain-general] but also the underlying heuristic principles, can facilitate conceptual understanding" (p. 784, interestingly this paper was submitted, reviewed, and accepted for publication when R. Duschl was the editor of the journal *Science Education*). For example, the rivalry between the Thomson and Rutherford models of the atom (cf. Chaps. 3 and 4) constitutes a heuristic principle as it helps to structure inquiry and understand an important facet of NOS, namely, scientific progress is characterized by competition among rival theories. Table 3.1 provides a summary of all the historical episodes and the resulting heuristic principles, discussed in the previous section. This clearly shows that both domain-general and domain-specific aspects of the nature of science are not dichotomous but rather integrated and are essential if we want to understand "science in the making" (Niaz 2012a). In other words, if we discuss science content (domain-specific), then it is helpful to introduce domain-general aspects and hence the integration. Chapter 8 provides further details as to how the two NOS aspects can be integrated in the context of "performance expectations" suggested by the Next Generation Science Standards (NRC 2013). See the next section for an illustration of the integrated view.

Introducing the Integrated View of the Nature of Science to In-Service Teachers

Inclusion of this section at this stage is important in order to illustrate how the questions suggested by the family resemblance view (Irzik and Nola 2011) of the nature of science (NOS) can be included in the classroom as part of teaching about the domain-specific aspects of NOS. The objective of this study is to facilitate science teacher's knowledge of NOS, based on the following research question: How can we provide opportunities to teachers' for understanding the scientific method, objectivity, and scientific development?

Method

This study is based on the participation of 12 in-service teachers who had enrolled for the following required course: "Science, Technology, Ethics, and Creativity in Research," as part of their doctoral degree program in materials science, at a major university in Venezuela. The course was subdivided into the following six topics:

(1) Introduction; (2) Scientific development: A history and philosophy of science perspective; (3) Methods and creativity in research; (4) Structure of a scientific publication; and (5) Review of research and the use of the Internet; (6) Evaluation of the efficiency of research based on good lab practice. The study reported here was based on topic 2 and the author was the instructor. Participants' undergraduate degree ("Licenciatura" or equivalent) was in the following fields: Chemistry = 1, Physics = 6, Petroleum engineer = 1, Chemical engineer = 2, and Systems engineer = 2. The age of the participants ranged from 25 to 40 years old, and the teaching experience ranged from 5 to 15 years (female = 2, male = 10). All participants were working in universities and colleges and most of them were giving freshman courses. Some of them already had a master's degree and publications in international research journals in their respective fields. A few of the participants had basic knowledge of the work of Popper, Kuhn, Lakatos, Giere, and of other philosophers of science. This study has benefitted from a previous study with participants having a similar background (cf. Niaz 2012a, Chap. 6).

Course Outline

All participants were provided with the following course outline, one month before they met with the instructor:

> This course topic is designed to facilitate an understanding of scientific development based on a history and philosophy of science perspective (Popper, Kuhn, Lakatos, Feyerabend, Laudan, Cartwright, Giere and others). Among others, the following experiments will be discussed:
>
> 1. The Michelson–Morley experiment to determine the velocity of the earth with respect to the ether (1887)
> 2. J.J. Thomson's cathode ray experiments (1897)
> 3. E. Rutherford's alpha-particle experiments (1911)
> 4. N. Bohr's understanding of experiments that determined spectra of elements (1913)
> 5. R. Millikan's (1910–1925) oil drop experiment to determine the elementary electrical charge (electron)
> 6. Experiments related to photoelectric effect as evidence for Einstein's hypothesis of light quanta (1916)
> 7. The bending of light in the 1919 eclipse experiments as evidence for Einstein's theory of general relativity
> 8. Martin Perl's experiments to isolate "quarks" at the Stanford Linear Accelerator Center (SLAC) (Perl was awarded the physics Nobel Prize in 1995)
>
> A history and philosophy of science perspective reveals that in these experiments (and others), scientific development is characterized by: (a) Experiments are important; however, their interpretations are even more important. (b) Rivalries and conflicts between scientists have repercussions in the development of their theories. (c) At times, the same data can be explained/interpreted through different models and theories. (d) Scientific theories are tentative. (e) Scientists do not inevitably abandon their theories on finding anomalous data. (f) The scientific method does not necessarily play an important part in scientific research.
>
> Reference:
>
> Niaz, M. (2009). *Critical appraisal of physical science as a human enterprise: Dynamics of scientific progress*. Dordrecht: Springer (available in the local library).

Reading Material:

(a) Niaz, M. (2010). Are we teaching science as practiced by scientists? *American Journal of Physics, 78*(1), 5–6.
(b) Niaz, M et al. (2010b). Leon Cooper's perspective on teaching science: An interview study. *Science & Education, 19*(1), 39–54. (Cooper was awarded the physics Nobel Prize in 1972, for his contribution to the theory of superconductivity.)

Note: All participants were provided with copies of the two articles in the reading material which were compulsory reading. The book was not required reading; however, a copy was available in the university library.

Course Organization and Activities

Participants had the opportunity to study the reading materials during the first four weeks of the course and also communicated among themselves and with the instructor through emails. At the end of the fifth week, participants met with the instructor for a 4-h session, which included the following activities: (a) a formal PowerPoint presentation by the instructor, based on 8 experiments referred to in the course outline (90 min); (b) an interactive discussion, based on the presentation and other aspects of the course (30 min); (c) recess (30 min); and (d) an interactive discussion with respect to various aspects of the course material (90 min). After another four weeks participants met with the instructor for a 4-h session, which included the following activities: (a) Each participant was given approximately 5 min to present his comments or criticisms of the different aspects of the course material (90 min); (b) recess (30 min); (c) general discussion (30 min); and (d) evaluation using a 3-item questionnaire (90 min). This evaluation was open-book and participants could use any materials or personal notes that could be of help. The questionnaire is presented in the next section.

Evaluation

All participants responded to the following 3-item questionnaire, which formed part of their formal evaluation for this course:

1. Many students, professors, science textbooks and methodology courses emphasize the scientific method. Do you think the scientific method always plays a primordial role in scientific development? (Primordial in this context means of fundamental importance.)
2. Mario Vargas Llosa (Nobel Laureate in literature 2010) in his Nobel Prize acceptance speech declared: "Literature is a false representation of life that nevertheless helps us to understand life better, to orient ourselves in the labyrinth where we are born, pass by, and die" (Vargas Llosa 2010). Let us compare this with the analogy drawn by Leon Cooper (Nobel Laureate in physics 1972), between a style of painting (impressionism) and scientific progress: "I believe, in some

ways, the scientist can be compared to the painter. The impressionists, for example, were accused of not being able to see things as they are. But, having imposed their way of viewing—their vision of the world—it has become a cliché now to see things as the impressionists did" (reproduced in Niaz et al. 2010b, p. 48). Let us summarize, according to:

Vargas Llosa: Literature is a false representation of life.
Cooper: The impressionists were accused of not seeing the things as they are.

(a) Do you think there is some coincidence or discrepancy between the opinions of Vargas Llosa and Cooper?
(b) In your opinion, what are the implications of these citations for understanding scientific development?

3. Many scientists, science textbook authors, and professors believe that science is "objective." If we accept this perspective, Newton's laws constitute the best example of objectivity in science. Nevertheless, at the beginning of the twentieth century, Einstein's theories of relativity (special and general) questioned Newton's laws. Accordingly, do you think that Newton's laws are false and consequently that he was not "objective"?

At this stage it is important to note that: (a) The three items of the questionnaire were closely related to the content of the course and classroom discussions and were not necessarily designed for this research report; (b) Extreme care was taken to guard the anonymity of the participants while quoting from their written responses. Some responses were shortened to avoid repetition, while maintaining the essential argument; (c) Participants were explicitly told that there were no right or wrong responses in this evaluation and they were free to express their opinions; (d) Both the reading material and classroom discussions provided participants with an overview of "science in the making" (Niaz 2012a) and the underlying NOS aspects (Niaz 2001a); (e) Due to their physical science background, all participants were aware of the context of the reading material and the questionnaire (domain-specific NOS); (f) Participants were exposed to a history and philosophy of science perspective, in order to understand scientific progress (domain-general NOS). However, they were told that they did not have to necessarily agree with this perspective; (g) It is plausible to suggest that this study is based on a "reflective, explicit, activity-based approach" (cf. Akerson, Abd-El-Khalick and Lederman 2000; Khishfe and Lederman 2007; Smith and Scharmann 2008).

Results and Discussion

Item 1 (Scientific Method)

Based on the participants' responses, the following three classifications were generated:

(a) Scientific method is primordial ($n=4$).
Four participants considered the scientific method to be indispensable and hence primordial for scientific progress and the following are two examples:

> The scientific method is based on observation, experimentation, enunciation of laws and theories and the confirmation of these laws and theories … The scientific method helps to reproduce a particular experiment with the objective of demonstrating that all scientific propositions are susceptible to falsification. Based on this further experiments can be designed which give different results from those expected initially. This is the essence of scientific development. (Participant #1)

> The scientific method plays an important and primordial role in scientific development based on the history and its applied philosophy, as certainly the data, theories and conjectures in the research must follow the scientific method. Observation, experimentation, enunciation of laws and theories and their confirmation, precisely constitutes development of science … Science is considered to be a reflexive and systematic procedure and in this process we use the scientific method, in order to discover or interpret the facts, phenomena, relations and laws in a particular part of the reality. (Participant #6)

(b) Scientific method is partially primordial ($n=2$).
Two participants considered the scientific method to be partially primordial and the following are two examples:

> It plays a primordial role but not completely. It is a part of the jigsaw puzzle that helps to orient scientific development in a formal manner. However, why do I consider it to be partially primordial? The answer is that often researchers omit certain results and manipulate the scientific method in order to demonstrate what they want and not what they originally wanted to find … In my opinion, the scientific method is part of the research but does not represent it completely. It does, however, serve as a guide and provides us with an idea as to how the experiments and the results need to be developed. (Participant #7)

> The scientific method will always be and should be used as a base for the development of science. Nevertheless, this method cannot be used in a strict sense or as an inquisitive in order to sustain new theories or findings. The actual vision of science comprises of a conglomerate of almost infinite hypotheses, theories and conjectures—and many of these lead to inconsistencies … We must find a point of inflection in which the scientific method is allowed to assimilate intuitions and inconsistencies for a more universal development of science. (Participant #11)

(c) Scientific method is not primordial ($n=6$).
Six participants considered the scientific method to be not primordial and the following are three examples:

> Scientific method is an implement used by every investigator to obtain information related to a problem both in the natural and social sciences. However, in the history of the natural sciences it has been found that the scientific method as understood in the scientific community has not been followed in a strict and rigorous manner. One example is the discovery of charge of the electron (1.602×10^{-19} C), based on the "oil drop experiment." There is evidence that Millikan discarded data obtained in his experiment, which means that he did not follow or respect the scientific method rigorously and still his findings are to this day accepted by the scientific community. (Participant #2)

> Ironically, science does not always develop with the scientific method … In their origin and development investigations have a part that is based on speculation. In the history of science, the great scientists worked very hard to reach their goals. However, in many cases

> the origin of their work is based on the ideas and presuppositions of previous theories and experiences, which in turn had some degree of speculation and prior beliefs. In other words, their "objective" results in some cases were born based on a subjective idea. (Participant #5)

> The scientific method is a scheme or a series of steps required for the development of an experiment. However, this is not followed strictly or a priori … development of science can follow many and diverse forms … The freedom of a scientist cannot be curtailed to a degree that makes him follow only one method—there must always be some indication of an idea, a hypothesis—leading to multiple ways of finding expected or unforeseen results that may generate great discoveries. (Participant #10)

Comments

Participants who considered the scientific method as primordial were quite explicit about the importance of the sequence of steps that help the scientist to obtain experimental data, which in turn helps to formulate scientific theories or laws. One participant even considered that the scientific method helps to demonstrate that "all scientific propositions are susceptible to falsification." Partially primordial responses recognized the importance of the sequence of steps but considered them to be incomplete and suggested the inclusion of scientists' inconsistencies, intuitions, and conjectures. Such changes may even constitute a "point of inflection" in the development and use of the scientific method. Participants who explicitly considered the scientific method not to be primordial referred to actual examples from the history of science in which the method was not followed (e.g., the oil drop experiment). However, even these participants recognized the importance of the scientific method, and they also considered that it might curtail the scientists' ability to use diverse methods depending on the needs of a particular experiment. This clearly shows that at least some participants based their evaluation of a domain-general NOS aspect (scientific method) on science content that was discussed in class (domain-specific NOS).

Item 2 (Vargas Llosa Versus Cooper)

Before analyzing participants' responses to this item, it is important to consider the context in which Leon Cooper made his comments (see Item 2 above). In his textbook, Cooper (1992) stated: "It seems to me that important physics, as important painting, imposes the *vision of the scientist/artist* on the raw data, in principle available to everyone. A generation or two later the world appears to us as that vision" (p. xii, emphasis added). In their interview study, Niaz et al. (2010b) asked Cooper: "What is your understanding of 'the vision of the scientist'?" (p. 47). Cooper's statement as reproduced in Item 2 is a response to this question.

(a) *Participants' responses to Item* 2a (Do you think there is some coincidence or discrepancy between the opinions of Vargas Llosa and Cooper?)

Interestingly, all 12 participants considered that there is a coincidence in the opinions of the two Nobel Laureates: Mario Vargas Llosa (a novelist) and Leon Cooper (a physicist). The following are some examples of participants' responses:

> Both [Vargas Llosa and Cooper] coincide that we need to slightly distance ourselves from the reality in order to understand it better. In other words, we need to momentarily go out of the frame of reference of an entirely objective world and submerge subjectively with imagination and creativity in a world with a different vision. (Participant #5)

> Despite the coincidence, both represent their ways of understanding and orienting life in different ways, which finally approximates to a "magical realism", that at some moment in life is real. (Participant #6)

> If we observe the background it can be noted that neither of them refers to an absolute truth. What they do refer to is a perception of something that is waiting, something that is being sought, something that may help to add another stepping stone in our "true" understanding of the world. (Participant #7)

> Both consider that in science and literature, creations are circumstantial. The events that lead to scientific theories or literary works are temporal and may change over a period of time. However, this sets the stage for later works and consequently writers and scientists must be free to create and express in their different disciplines, what they perceive of the world without being pigeonholed in strict methodologies. (Participant #10)

Comments

These responses clearly show that in order to create a new vision of the world, we need to go beyond our present frame of reference that requires imagination and creativity. At the same time, both the writer and the scientist are always looking for "something" that they do not know how to find and still keep persevering in this never-ending quest for yet another stepping stone. Again, this "journey" becomes easier if we are not forced to follow certain predetermined sequences of research methodologies. Magical realism refers to a style of literature that departs from reality in order to distance us from known and existing forms of thinking and in the end help us to understand our present predicaments. Interestingly, Mario Vargas Llosa and Gabriel García Márquez (another Nobel Laureate) are considered to be the prime exponents of this style of writing. García Marquez's novel *Cien Años de Soledad* (100 years of solitude) is considered a world classic and has been translated into many languages. Interestingly, some participants explicitly referred to the following domain-general NOS aspects, without being asked to do so: relationship between subjectivity and objectivity, role of imagination and creativity, and absolute truth and methodology.

(b) *Participants' responses to Item* 2b (In your opinion, what are the implications of these citations for understanding scientific development?)

In this part of the Item, participants were asked to give their opinion with respect to the possible implications of the two citations (Vargas Llosa and Cooper) for scientific development. The following are some examples of participants' responses:

> The natural sciences have developed thanks to the different visions, correct or mistaken, of many investigators who have dedicated themselves to understand the physical world. In other words, everybody does not agree with respect to the way we understand the world. All the scientific progress of our world is in the final analysis an accumulation of different representations, and *for different reasons some representations come to impose themselves on others and later are accepted by the scientific community*. (Participant #2, italics added)

> The development of science depends on the creativity, audacity and the hard work of the scientists. We should not isolate ourselves in a box of objective knowledge, like machines for the production and replication of experiments. On the contrary, we *need to innovate by putting into practice our creativity*, just like the scientists that have been recognized historically. Only in this way we shall be able to understand scientific development and what is more contribute towards its growth. (Participant #5, italics added)

> Both Cooper and Vargas Llosa are trying to explain that there are two types of thinking that permit us to make sense of the world, despite the disapproval of many. In order to have a different scientific development, we have to break the dogmas of the previous scientists. *We do not have to do science with just one perspective but rather formulate new methods* which may collide with the traditional way of doing science. (Participant #8, italics added)

> Thinking differently, permits the *plurality of thinking* and this helps scientists to facilitate development of science and society. (Participant #9, italics added)

> What is known and observed at a given moment in time is not the *absolute truth*, but rather conditioned by many factors that can vary with time. This leads to the emergence of new conditions and *alternative points of view* that may complement or change entirely a theory or what was previously conceived as true. (Participant #10, italics added)

> These citations [Cooper and Vargas Llosa] show that in the development of science there will always be *a confrontation between those who claim to "have the truth" and those who struggle to find it*. Those who are struggling, on the one hand fear that they may be mistaken and on the other hand they have to face the criticisms of those who developed the currently accepted theories. (Participant #11, italics added)

> As the world is changing, if we anchor our thinking in the old ways of doing science, we shall never make any contribution to the world in which we live. Scientific development involves having new ideas, that *a model is not absolute*, acceptance of a theory leads to its critical evaluation and finally it is changed. (Participant #12, italics added)

Comments

Most participants were quite enthusiastic about the implications of the views of Vargas Llosa and Cooper. Interestingly, if we analyze closely the responses reproduced above, these almost constitute an agenda for introducing the nature of science in the classroom, based on the following aspects (domain-general NOS): (a) for different reasons some representations come to impose themselves on others and later are accepted by the scientific community; (b) the need to innovate by putting into practice our creativity; (c) we do not have to do science with just one perspective but rather formulate new methods; (d) plurality of thinking; (e) alternative points of view; and (f) a model is not absolute. These aspects of nature of science form part of many reform documents and would be accepted by many science educators (Niaz 2012a). Participant #10 raised a very important issue, namely, that *science does not look for absolute truths*. Participant #11 went even further by pointing out a confrontation, or even perhaps a struggle, between those who claim to already have the "truth" and those who are struggling to find alternative points of view, perhaps the "new truth." These are important issues and are being discussed by contemporary philosophers of science. According to Giere (2006a, b), scientific theories are not necessarily "true" or "false" but rather provide us the means to understand the world and are in a continual process of change (for details see Chap. 2). Participants #5 and #12 explicitly referred to the need for studying science (e.g., replication of experiments) and also for contributing toward the progress of science. Again, this is an important issue, as very few reform documents recognize the importance of stimulating students to contribute toward the discovery of "new truths." To make matters worse, in many parts of the world, science students are not only expected to replicate experiments but rather learn to crank a well-oiled machine of discovery (cf. Calver 2013, also Chap. 2 for details). Participant #9 has referred to the importance of *plurality of thinking* in the progress of science. Again this is an important issue being discussed by philosophers of science. Giere (2006a, b) has referred to this as *methodological pluralism* and its implications will be discussed in Chap. 8.

Item 3 (Newton Versus Einstein)

The background to this item is provided by Giere's (2006a) critique of those scientists and philosophers of science who consider that what drives scientists onward is that there are *truths out there to be discovered* and that such philosophical positions can be considered as "objectivist realism" (see Chap. 2 for details). Of the 12 participants, 10 stated that Newton's laws were not false and that he was "objective" in the formulation of his laws, and the following are some examples:

> First it is important to recognize that Newton molded his vision of the material world based on the law of universal gravitation, thanks to the work of scientists such as T. Brahe, N. Copernicus, J. Kepler, and G. Galilei. Was Newton objective in the formulation of his theory? He thought that he was and many believed that his vision was the last word with respect to this problem. However, Einstein demonstrated with his theory of relativity that Newton was not sufficiently objective as his theory could not explain certain phenomena that the theory of relativity could. But thanks to Newton, Einstein could see beyond Newton. Are Newton's laws false? In physics it is known that these laws are not fulfilled in the context of Einstein's physics and consequently are not objective in this context. Nevertheless, these days Newton's laws continue to be applied, and consequently, I think that in a certain sense these laws have "some degree of truth" in their natural context of application. Was Einstein objective? Until now history tells us that he was. For how long? We still do not know. (Participant #2)

> Newton's laws are not false. As in any other work of investigation there are always conflicts, counter-arguments and different opinions. Einstein questioned Newton's laws as they could not be applied exactly to all the particles in space and subatomic particles that behave differently. Newton was objective in his conclusions, although we need to know the *origin of his ideas, which were perhaps part of a dream or a speculation*. However, it is important to note that Newton's laws can be applied in classical mechanics with a fairly good approxi-

> mation and objective manner, especially when the velocity of the particles is much less than that of light. (Participant #5, italics added)

> Newton's laws are not false. I believe that at that time in history his laws revolutionized ways of thinking and helped to explain many phenomena that could not be understood earlier. Similarly, he was objective as his laws helped to solve many problems … In the quest to understand further, Einstein based on the theory of relativity questioned Newton's laws. Now, in the twenty-first century someone with a new theory would do the same to what Einstein did to Newton's theory. *The "truth" has no end and it is "true" until someone discovers a new theory that is accepted by the scientific community.* (Participant #9, italics added)

Two participants considered that Newton's laws were false, although they could still be considered as objective and the following is an example:

> From our actual state of the knowledge, I do believe that laws of Newton are false as he did not consider movement of objects at high velocities approaching that of light. Einstein explained that Newton's laws can be applied only to Euclidean systems (macroscopic world). Newton's laws, however, can be considered as objective as they are still used to explain some phenomena. Furthermore, laws of Newton contributed to Einstein's elaboration of the theory of relativity. (Participant #3)

Comments

Most philosophers of science (including Duhem, Giere, Kuhn, Lakatos, and Laudan) would agree that if a scientific theory is replaced by another with greater explanatory power, it does not mean that the previous theory was either false or that its author was not being "objective." This is the dilemma faced by the participants in this item. In other words, Newton's laws when first proposed in the seventeenth century were "true" for that time (actually for more than 200 years), and he was as "objective" as one could possibly expect a scientist to be. Consequently, the solution to the dilemma lies in recognizing that both Newton and Einstein were being "objective" and provided theories that varied in their explanatory power in certain domains (e.g., Einstein explained better the behavior of particles approaching the velocity of light).

With this background, it is easier to understand the responses provided by the participants of this study. It seems that a majority (10 out of 12) of the participants had a fairly good understanding of the role of "truth" of a theory and consequently the "objectivity" of the scientist. Following Giere (1999, 2006a, b) scientific theories are not "true" or "false," and similarly the role of the scientist is more perspectival rather than "objective" (see Chap. 2 for details). However, a review of the literature (Blanco and Niaz 1997, 1998) shows that if a scientific theory is replaced by a more successful theory, students attribute this to the fact that the earlier theory was false and consequently its author was not entirely "objective." This was the perspective adopted by the two participants in this study who responded that after Einstein's theory of relativity, Newton's laws can be considered to be "false."

Interestingly, these participants did recognize that Newton was "objective" in the elaboration of his laws.

At this stage it would be interesting to have a closer look at the responses provided by some participants. Participant #2 tried to understand Newton's contribution in a historical context by recognizing the work of Brahe, Copernicus, Kepler, and Galileo, which is a sound approach. However, this participant was clearly struggling to understand the dilemma, as she/he asked, "Was Newton objective in the formulation of his theory?," and again responded in a historical context by pointing out that "many believed that his vision was the last word with respect to this problem." Next this participant reminded us that "But thanks to Newton, Einstein could see beyond Newton," and this helped to respond to the question, "Are Newton's laws false?" Finally, this participant raised a thought-provoking question, "Was Einstein objective?" and responded laconically, "For how long?" In my opinion, this line of reasoning approximates to a metaphoric representation of a Newton–Einstein debate.

Participant #5 raised a very important issue with respect to the very "origin of Newton's ideas, which were perhaps part of a dream or a speculation." Indeed, a historical reconstruction of the origin of Newton's ideas is revealing with respect to scientific research methodology and shall be discussed later in the Conclusion section. Participant #9 clearly referred to the tentative nature of scientific theories by pointing out that a scientific theory is "true" until someone propounds a new theory. Once again, participants in this study went out of their way to understand the domain-specific NOS aspects (Newton and Einstein frameworks) within domain-general NOS aspects (absolute truth, objectivity).

Conclusion

Results obtained in this study reveal that given the necessary experience (domain-specific historical episodes), in-service teachers are quite receptive and willing to give up some of their well-ingrained aspects of an empiricist epistemology based on various NOS aspects (cf. Niaz 2012a, Chap. 4). The line of reasoning followed by Participant #2 based on the one hand on a historical reconstruction and at the same time interspersed with arguments and counterarguments is indeed the very essence of what we as science teachers need to facilitate.

The reference to the *origin of Newton's ideas* by Participant #5 was quite unexpected and did not form part of classroom discussions. This leads to the question: Did Newton formulate his laws based entirely on experimental observations? Most science textbooks and even curricula endorse this empiricist epistemology. A historical reconstruction of the origin of Newton's ideas reveals a different and a very interesting story (see Niaz 2009, Chap. 2 for details). If the answer to the question posed above is in the affirmative, then Newton should have been aware that charged bodies would not follow the law of gravitation (but he was not, as this law was discovered well after his death). According to Cartwright (1983), the law of universal gravitation and Coulomb's law interact to determine the final force. An inquisitive student may respond that perhaps Newton did consider the charge of the bodies and even collected experimental data. Insight from Giere (1999) can help to resolve this dilemma:

> Most of the laws of mechanics as understood by Newton, for example, would have to be understood as containing the proviso that none of the bodies in question is carrying a net charge while moving in a magnetic field. That is not a proviso that Newton himself could possibly have formulated, but it would have to be understood as being regularly invoked by physicists working a century or more later. (p. 91)

Similarly, according to Kuhn (1977), when Newton enunciated his theory in the late seventeenth century, only his third law could be directly investigated by experiment. Convincing demonstration of the second law had to await the development of Atwood's machine, almost a century after the appearance of Newton's *Principia.* Duhem (1914) suggested ways for testing Newton's first law of inertia, which specifies the behavior of those bodies which are under the influence of no impressed forces. However, no such body exists, as an observed body cannot be free of impressed forces (Losee 2001). Consequently, Newton's law of inertia cannot be a generalization about the observed motions of particular bodies.

In order to facilitate understanding, let us consider the following dilemma posed by Duhem (1914), a physical chemist–philosopher of science: "Does logic require our hypotheses to be simply experimental laws generalized by induction?" (p. 219). Most science textbooks and curricula would respond in the affirmative. Duhem (1914) himself responded, and even after 100 years, some teachers and even scientists may feel uneasy with his forthright approach:

> Now, we have recognized that it is impossible to construct a theory by a purely inductive method. Newton and Ampère [André-Marie Ampère 1775–1836] failed in this, and yet these two mathematicians had boasted of allowing nothing in their systems which was not drawn entirely from experiment. Therefore, we shall not be averse to admitting among the fundamental bases of our physics postulates not furnished by experiment. (p. 219)

Ampère is known for his theory of electrodynamics and in his treatise stated that some of the experiments on which the theory was based had still to be performed. Duhem (1914) finally concluded that the "Newtonian method," attractive as it may appear, was a dream.

Let us go back and look at the response of Participant #5, "… origin of his [Newton's] ideas, which were perhaps part of a dream or a speculation." The similarity between Duhem's views and that of the participant is striking indeed. However, given the previous training (traditional physical science empiricist epistemology), her/his response is quite thought-provoking. If we scrutinize closely the line of argument of Participant #5, it is plausible to suggest (parts of the original response in italics, followed within parentheses by my comments):

- *Newton's laws are not false*. (This follows from classroom discussions based on Giere's naturalism.)

- *As in any other work of investigation there are always conflicts, counter-arguments and different opinions.* (Again this was discussed in class in different historical episodes, e.g., Thomson–Rutherford and Millikan–Ehrenhaft controversies.)
- *Einstein questioned Newton's laws, as they could not be applied exactly to all the particles in space and subatomic particles that behave differently.* (This was discussed in class, and most physical science students are familiar with the limitations of Newtonian mechanics.)
- *Newton was objective in his conclusions.* (Again this was discussed in class in the context of how even if a theory is later considered to be false, the work of the scientist is still "objective.")
- *Although we need to know the origin of his ideas, which were perhaps part of a dream or a speculation.* (This is the intriguing part of the response as it was not discussed in class.)
- *However, it is important to note that Newton's laws can be applied in classical mechanics with a fairly good approximation and objective manner, especially when the velocity of the particles is much less than that of light.* (It seems that the participant reasoned that Newton did not know of subatomic particles and still his laws are approximately applicable, leading to the conclusion: the origin of at least some part of Newton's work lies in speculation.) At this stage it is important to recall that this participant might have consulted Niaz (2009) to make these statements, which discusses these issues and was available in the local library. Nevertheless, the context in which the participant referred to these issues within a sequence of arguments still constitutes a very novel approach.

Nature of Science in the Context of Chemistry/Science Education

Understanding the nature of science in science and chemistry education is a complex issue and has been the subject of considerable controversy in the science education literature (see earlier sections of this chapter for details). In this section, I will provide examples from the literature to show that if we want to understand the nature of science, some degree of integration of domain-general and domain-specific aspects is essential. In other words, there is no single way of understanding or teaching nature of science. Consequently, integration based on pluralism is almost an imperative for science educators. This section is based on the following aspects of nature of science:

(a) Nature of science in textbooks
(b) Students' and teachers' understanding of the nature of science
(c) Teaching the nature of science in the classroom

Nature of Science in Textbooks

In Chap. 3, I provided examples showing that the domain-general heuristic principles formulated by philosophers of science are themselves based on an in-depth (domain-specific) historical reconstruction of particular episodes in the history of science. Furthermore, I provided examples from research in science education to show that both the domain-general and domain-specific aspects of nature of science complement each other and are essential for facilitating students' conceptual understanding of the various topics of the science curriculum. Based on the same arguments, in this section, I would like to differentiate between two types of studies related to science textbooks:

(a) Domain-specific: These studies are based on a historical reconstruction of a given topic of the science curriculum. The following are some examples of such studies from the science education literature: quantum hypothesis (Brush 2000), photoelectric effect (Niaz et al. 2010a), periodic table (Brito et al. 2005), atomic structure (Niaz 1998b, 2000a; Justi and Gilbert 1999; Padilla and Furio-Mas 2008), and oil drop experiment (Niaz 2000a). For example, criteria used for evaluating the presentation of atomic structure in textbooks cannot be used for evaluating the periodic table—in this sense the criteria are domain-specific.
(b) Domain-general: These studies are based on a series of nature of science (NOS) theses, which are in turn based on a history and philosophy of science perspective. Such theses consider NOS to be empirical, tentative, inferential, creative, theory-driven, social, culturally embedded, and others. The following are some examples of such studies presented in the science education literature: Abd-El-Khalick, Waters, and Le (2008); Leite (2002); Niaz and Maza (2011); and Vesterinen et al. (2013).

It is helpful for students and teachers if authors of domain-specific studies draw conclusions with respect to the generality of the various NOS theses. Similarly, authors of domain-general studies could relate these dimensions with a particular context of the science curriculum. It would be very helpful if the tentative nature of science could be exemplified by the changing nature of atomic models, which form part of almost all science textbooks both at the high school and the introductory university level (viz., integrated view of nature of science). For example, most high school and introductory university textbooks deal with the atomic models of Dalton, Thomson, Rutherford, and Bohr. This means that some history is already there. However, what lacks are the controversies, the reasons why models change, and, thus, an explicit presentation of the tentative nature of science.

At this stage I would like to present two examples of studies that have evaluated the representation of nature of science in two contexts: (a) high school chemistry textbooks (Abd-El-Khalick et al. 2008) and (b) introductory university-level general chemistry textbooks (Niaz and Maza, 2011). Both can be considered as examples of domain-general studies.

Abd-El-Khalick, Waters, and Le (2008) have drawn attention to the importance of including the nature of science (NOS) in high school chemistry textbooks. These authors analyzed 14 textbooks (published in the USA, 1966–2005) including five "series" spanning one to four decades, with respect to the following NOS aspects: empirical, tentative, inferential, creative, theory-driven, myth of the scientific method, nature of scientific theories and laws, and the social and cultural embeddedness of science. Based on the scoring rubric designed for this study, all three authors analyzed all textbooks independently and attained an inter-rater agreement of 86 %. Results from this study revealed that high school chemistry textbooks fared poorly in their representation of NOS, which led the authors to conclude: "These trends are incommensurate with the discourse in national and international science education reform documents [AAAS 1989; NRC 1996] ..." (p. 835). Authors considered the following finding to be the most disturbing: all textbooks (except Toon et al. 1968) espoused the die-hard myth of the "scientific method" (p. 848). Interestingly, Niaz and Maza (2011) in a study designed for evaluating the nature of science found that Toon and Ellis (1978) had the highest score in a sample of 75 general chemistry textbooks. Abd-El-Khalick et al. (2008) refer to this as an "author effect" as compared to a "publisher effect." In other words, science educators could approach textbook authors with well-formulated and documented arguments in order to facilitate the inclusion of such facets of NOS and HPS in their textbooks.

Furthermore, in order to understand the significance of the tentative nature of science, consider the following example from a textbook that was considered to be an explicit, informed, and consistent representation:

> Even today, after a number of modifications, our model of the atom continues to undergo constant change as new evidence accumulates and new theories are developed. However, no matter how detailed a model of the atom becomes, it can never depict the true structure of the atom. It is important to avoid falling into the trap of taking models too literally. You must bear in mind their limitations and remember that all of them fall short of reality. (Toon et al. 1968, p. 7; reproduced in Abd-El-Khalick et al. 2008, p. 846)

Now let us compare this presentation to that of a textbook that was considered to be an implicit and not an informed representation of NOS:

> Models, like theories, are refined as new information is discovered ... certain facts in science always hold true. Such facts are labeled as scientific laws. (Tocci and Viehland 1996, p. 20; reproduced in Abd-El-Khalick et al. 2008, p. 846)

A major point of difference between the two presentations is that the one by Toon et al. (1968) sends out a clear message to the student. On the contrary, the presentation by Tocci and Viehland (1996) includes statements that convey conflicting messages about the same NOS aspect. A comparison of the two presentations clearly shows how the tentative nature of scientific knowledge becomes meaningful to the students only if it is immersed within a domain-specific context that forms part of the science curriculum (in this case atomic models and similarly other examples can be found). Examples from these two textbooks provide teachers an opportunity to reconsider Duschl and Grandy's (2013) advice with respect to the tentative nature of science and ask which of the two books they would recommend to their students.

To provide an example from a different cultural context, Vesterinen, Aksela, and Lavonen (2013) found that the tentative nature of science was the most common dimension of NOS emphasized in all the five upper secondary school textbooks published in Finland and Sweden. Some of the examples used to explain this dimension were the development of atomic models, discovery of the unknown elements, creation of the periodic table, and synthesis of new substances in drug discovery. Interestingly, the Swedish core curriculum explicitly mentions the tentative nature of scientific knowledge.

Niaz and Maza (2011) have analyzed the introductory chapter (or preface) of 75 general chemistry textbooks (published in the USA, 1965–2008). A review of the literature shows that textbook authors do not necessarily present a consistent NOS perspective in all the chapters of their book. In addition, the introductory chapter of a textbook can provide an overall NOS perspective of the author, to a fair degree. Consequently, this study is based on an analysis of the introductory chapters of general chemistry textbooks based on the following aspects/criteria: tentative nature of scientific theories; role of laws and theories; scientific method; observations are theory-laden; experimental evidence and rational arguments; competition between rival theories; different interpretations of the same experimental data; inconsistent foundations of scientific theories; and social and historic milieu. Textbooks were classified as Satisfactory (S), Mention (M), and No mention (N). The percentage of textbooks that were classified as N ranged from 44 % (Criterion 1, tentative nature) to 94.7 % (Criterion 8, inconsistent foundation). The percentage of textbooks that were classified as S ranged from 1.3 % (Criterion 2, role of laws and theories) to 17.3 % (Criterion 1, tentative nature).

These results show that although the presentation of NOS is not the major objective of these textbooks, some of them inevitably refer to the historical record and thus provide NOS guidelines for students, teachers, and future textbook authors. Some textbooks go into considerable detail to present the atomic models of Dalton, Thomson, Rutherford, Bohr, and the wave-mechanical one (see Chap. 4 for details with respect to the development of these models). However, the most important aspect of these presentations is that they explicitly do so in the context of the tentative nature of scientific theories (Criterion 1). This is a clear illustration of how the history of chemistry can facilitate the understanding of NOS. It is concluded that in most cases the history of chemistry is already "inside" chemistry, and in order to facilitate conceptual understanding, textbooks need to interpret the development of events within a NOS perspective (see Chap. 1).

The following is an example of a textbook that was classified in the study by Niaz and Maza (2011), as Satisfactory (S) on Criterion 2 (role of laws and theories):

> It is important to understand that a law, that correlates a series of observations, is essentially empirical; it only registers and summarizes in a concise manner the results of a great number of experiments … theories explain observations according to an imaginary framework, not directly observable, and predict what has not been observed so far. For example, a law that we attribute to Boyle affirms that at low pressures the volume of a gas is inversely proportional to the pressure exerted on the vessel. A theory suggests that Boyle's law is

obeyed as particles (molecules) of gaseous material are far from each other and can easily approximate in order to increase the pressure, and can draw apart if the pressure is decreased. The law is observed directly, whereas the theory must always remain as a possible explanation, until the molecules of the gases can be observed directly. (Gray and Haight 1969, pp. 1–3; reproduced in Niaz and Maza 2011, p. 15)

The presentation by Gray and Haight (1969) comes quite close to how philosophers of science generally differentiate between theories and laws. For example, Losee (2001) differentiates between laws and theories in similar terms and provides examples of laws of nature, such as Boyle's law and Galileo's laws of free fall.

In order to differentiate between a textbook that was classified as Satisfactory (S) from the one that was classified as No mention (N) on this criterion, consider the following example:

Often a large number of related scientific facts can be summarized into broad, sweeping statements called **natural, or scientific, laws**. The law of gravity is a classic example of a natural law. This law—all bodies in the universe have an attraction for all other bodies that is directly proportional to the product of their masses and inversely related to the square of their separation distance—summarizes in one sweeping statement an enormous number of facts … Such a natural law can be established in our minds only by inductive reasoning; that is, you conclude that the law applies to all possible cases, since it applies in all of the cases studied or observed." (Joesten et al. 1991, p. 6, emphasis in original; reproduced in Niaz and Maza 2011, p. 15)

Now, let us compare this textbook presentation of Newton's law of gravitation, to that of Nancy Cartwright (1983, p. 57), a philosopher of science: No charged objects will behave just as the law of universal gravitation says; and any massive objects will constitute a counterexample to Coulomb's law. Consequently, these two laws are not true as generally understood by textbooks.

The following is an example of a textbook that was classified on Criterion 7 (different interpretations of the same experimental data) as Satisfactory (S) in the study by Niaz and Maz (2011):

Humans have always been fascinated by the heavens, by the behavior of the sun by day and the stars by night … the basic *observations* of these events have remained the same over the past 4000 years. However, our *interpretations* of the events have changed dramatically. For example, about 2000 B.C. the Egyptians postulated that the sun was a boat inhabited by the god Ra, who daily sailed across the sky … Eudoxus, born in 400 B.C. … imagined the earth as fixed, with the planets attached to a nested set of transparent spheres that moved at different rates around the earth … Five hundred year later, Ptolemy, a Greek scholar, worked out a plan more complex than that of Eudoxus, in which the planets were attached to the edges of spheres that "rolled around" the spheres … in 1543, a polish cleric, Nicolas Copernicus, postulated that the earth was only one of the planets, all of which revolved around the sun … Kepler postulated elliptical rather than circular orbits for the planets in order to account more completely for their observed motions. Kepler's hypotheses were in turn further refined 36 years after his death by Isaac Newton, who recognized that the concept of gravitation could account for the positions and motions of the planets … Einstein … showed that Newton's mechanics was a special case of a much more general model …. (Zumdahl 1993, p. 6, original italics)

Zumdahl (1993) has provided a good overview of how our *observations* related to the heavenly bodies (a topic of interest to most students) have been *interpreted* differently for almost 4000 years (2000 BC to 2000 AD), through the contributions of scholars belonging to different periods of time and cultures (Egyptians, Eudoxus, Ptolemy, Copernicus, Kepler, Newton, Einstein). This may encourage a student to ask: What is next? Again, this provides a good example of how NOS aspects can be incorporated into the textbooks. This discussion leads to another important issue: If observations can have varying interpretations, based on different theories and models which lead to controversies, can we conclude that this undermines the objective nature of science? To respond to this question, we consulted Leon Cooper, who responded in the following terms: "Observations can have varying interpretations, but this does not undermine the objective nature of science. It's somewhat ironic that what we like to call the meaning of a theory, its interpretation, is what changes. Think, for example, of the very different views of the world provided by quantum theory, general relativity and Newtonian theory" (reproduced in Niaz et al. 2010b). This clearly shows the importance of alternative interpretations of observations for science education, and the presentation of Zumdahl (a chemistry textbook author), quite similar to that of Leon Cooper (Nobel Laureate in physics), provides a good example.

Another example from the study by Niaz and Maza (2011) deals with Criterion 9 (social and historic milieu), and the following presentation was classified as Satisfactory (S):

> The development of scientific theories does not always happen easily, quickly, or smoothly. Evolution of thought takes time. The modern view of the solar system, for example, took thousands of years and countless astronomical observations to develop. At times, new ideas meet significant resistance. The famous Italian scientist Galileo Galilei (1564–1642) was forced by church authorities to retract his views that Earth moved around the sun … In the early 1900s, Marie Curie, a Polish- born French scientist, was a pioneer in the newly discovered field of radioactivity. Despite her many honors, including two Nobel prizes, she was never elected to the French Academy of Sciences. Apparently she was slighted because she was Polish born and a woman. In the 1950s, Linus Pauling, an American chemist, had his passport restricted by the government and was not allowed to travel out of the United States. In the 1970s and 1980s, Andrei Sakharov, a Russian physicist, was exiled to a small Russian city and not allowed to talk with other scientists. Both Pauling and Sakharov were punished for speaking against the development of nuclear weapons … Recently, the Catholic Church admitted that Galileo was treated unfairly in the 1600s, and Marie Curie's remains were moved to an honorary grave in the Pantheon of Paris 60 years after her death." (Dickson 2000, p. 6)

Discussion of such episodes from the history of science can provide students with an opportunity to understand the complexity of the scientific enterprise and appreciate that scientists are subject to the social and political norms prevalent at a given period of time (cf. Cooper 1992 reproduced in Niaz et al. 2010b; Longino 1990; Machamer and Wolters 2004). Indeed, the examples of Galileo, Marie Curie, and many others can help students to understand that "… both rationality and objectivity come in degrees and that the task of good science is to increase these degrees as far as possible" (Machamer and Wolters 2004, p. 9).

At this stage I would like to refer to what was suggested at the beginning of this section, namely, that authors of domain-general studies could relate these dimensions with a particular context of the science (chemistry) curriculum, thus facilitating integration. The following are some examples provided by Niaz and Maza (2011) and research in science education, for each of the nine criteria that they evaluated:

Criterion 1 (tentative nature of scientific theories):

Atomic models of Dalton, Thomson, Rutherford, Bohr, and wave-mechanical

Criterion 2 (role of laws and theories):

Boyle's laws can be observed, whereas the kinetic–molecular theory of gases is a possible explanation.

Criterion 3 (scientific method is an idealization of real science):

In the oil drop experiment did Millikan follow the scientific method? This could be the subject of considerable debate in the classroom.

Criterion 4 (theory-ladenness of observations, namely, what we observe is influenced by our theoretical frameworks):

(i) Thomson's rejection of Rutherford's interpretation of the alpha-particle experiments was based on his presupposition, namely, the "plum-pudding" model of the atom.
(ii) Millikan's interpretation of the photoelectric effect was based on his presupposition, namely, the classical wave theory of light.

Criterion 5 (role of insight, imagination, and creativity in science):

(i) Discovery of the structure of DNA
(ii) Bohr's postulation of the "quantum of action" to explain the paradoxical stability of the Rutherford model of the atom
(iii) Mendeleev's postulation of the periodic law

Criterion 6 (role of rival theories which leads to competition and controversies):

(i) Origin and development of the quantum theory (Bohr's Copenhagen interpretation and Bohm's "hidden variables")
(ii) Pauling's valence bond and Mulliken's molecular orbital theories to explain chemical bonding

Criterion 7 (alternative interpretations of experimental data):

(i) Understanding heavenly bodies (from the Egyptians, Greeks, Copernicus, Kepler, Newton, and Einstein)
(ii) Thomson's hypothesis of single-scattering and Rutherford's hypothesis of compound scattering to explain alpha-particle experiments
(iii) Millikan's and Ehrenhaft's hypotheses to explain data from the oil drop experiment

Criterion 8 (inconsistent nature of scientific theories):

(i) Phlogiston theory. According to Musgrave (1976), as early as 1630, it was a common knowledge that metallic oxides weighed more than the metals, and hence phlogiston theory was "born refuted."
(ii) Bohr's postulation of the "quantum of action."

Criterion 9 (social and historic milieu):

(i) According to Cooper (reproduced in Niaz et al. 2010b), due to the milieu of the time, Michelson–Morley experiment would not have made sense in the days of Copernicus. This has implications for science education if you teach about an experiment without providing the historical details of how and why it was done.
(ii) Difficulties faced by Copernicus and Galileo with the church authorities.
(iii) A. Sakharov and L. Pauling had difficulties in pursuing their work due to the opposition of the Soviet and US governments, respectively, in the twentieth century.
(iv) Explanation of the structure of oxygen molecule in valence bond and molecular orbital theories (Shaik and Hiberty 2008).

For further details with respect to the presentation of nature of science in science textbooks, see Niaz (2014a).

Students' and Teachers' Understanding of the Nature of Science

Students often believe that science is a collection of facts and that the best way to learn science is to memorize those facts (Linn et al. 1991). The degree to which students' conceptions of nature of science are influenced by their teachers and textbooks is the subject of considerable research. Such influence is mediated by a complex set of factors, such as curriculum constraints, administrative policies, and teachers' conceptualization of learning (Lederman 1992; Niaz 2011, 2012a). Teaching about NOS also contributes to developing scientific literacy, and that can be successful only to the extent that science finds a niche in the cognitive and cultural milieu of students (Cobern et al. 1999).

Niaz (2012a, Chap. 3) has reviewed research on students' and teachers' understandings of the nature of science within a history and philosophy of science perspective. This review is based on 94 articles published in the period 2004–2008 and has drawn on the following major science education journals: *International Journal of Science Education* ($n=34$), *Journal of Research in Science Teaching* ($n=28$), and *Science Education* ($n=32$). The following qualitative criteria were used for selecting the articles: title of the article, abstract, keywords, theoretical rationale, method, conclusion, and references. An article was selected only if it had a direct bearing on some NOS aspect within a history and philosophy of science perspective. After having selected the studies for review, it became apparent that these could be classified in the following two broad categories:

1. Epistemological beliefs of students and teachers with respect to the nature of science ($n=60$). Based on the type of study, this category was further subdivided into the following seven subcategories:
   (a) Relationship between students' and teachers' epistemological beliefs ($n=27$)
   (b) Myth of the scientific method ($n=3$)
   (c) Children's scientific reasoning ($n=4$)
   (d) Scientists' views of nature of science ($n=9$)
   (e) Nature of science and science curriculum ($n=10$)
   (f) Nature of science and students' laboratory practice ($n=6$)
   (g) Science exhibitions as a means to understanding the nature of science ($n=1$)
2. Facilitating students' and teachers' understanding of the nature of science (NOS), based on topics that are already in the science curriculum ($n=34$). Based on the type of study, this category was further subdivided into the following six subcategories:
   (a) Role of argumentation ($n=9$)
   (b) Explicit and reflective vs. implicit inquiry-oriented instruction ($n=11$)
   (c) Use of NOS-enriched instructional materials ($n=7$)
   (d) Use of history-based instructional materials ($n=3$)
   (e) Use of technology-based historical materials ($n=2$)
   (f) Use of science apprenticeship programs ($n=2$)

It is important to note that both the categories and subcategories were not generated a priori, but were based on the type of research reported in the three selected journals. In other words, issues explored in this chapter emerged from the review of the literature and not selected by the author.

Distribution with respect to the country (where the study was conducted, in the case of more than one author) was the following: USA=51, Taiwan=9, UK=7, Turkey=5, Canada=4, Australia=2, France=2, Israel=2, Korea=2, Lebanon=2, New Zealand=2, Argentina=1, Estonia=1, Hong Kong=1, Netherlands=1, Sweden=1, and Zimbabwe=1. This clearly shows that a major part of the work related to HPS and NOS was published in the USA, with smaller contributions from various other countries. All these studies emphasized some aspect of history and philosophy of science and its implications for science education. Furthermore, articles based on textbooks with a HPS perspective were not included in this study. In a previous study based on articles from these three journals over a five-year period (1998–2002), Tsai and Wen (2005) found 68 articles related to *history, philosophy, epistemology, and nature of science* (out of a total of 802 articles, 8.5 %). In a follow-up study, Lee, Wu, and Tsai (2009) found 71 articles related to this topic in the same three journals in the period 2003–2007. This shows the continued interest in this area of research. Some of the salient aspects of research reviewed in this study (Niaz 2012a) are the following:

(a) It is interesting to study how research in NOS-related areas has evolved (longitudinal aspect) as compared to earlier studies. However, the following aspect is noteworthy: In his review, Lederman (1992) had noted, "... both groups [students & teachers] have been berated for not understanding that scientific knowledge is necessarily tentative ..." (p. 352). Research reviewed in this study shows that despite some improvement, the tentative nature of scientific knowledge continues to be one of the most difficult NOS aspects in most parts of the world.
(b) As compared to previous research, this study reveals that instead of evaluating NOS views, there is more concern for enhancing students' and teachers' NOS and HPS understanding. Furthermore, there seems to be some consensus that instead of implicit, we need explicit and reflective teaching strategies embedded in HPS- and NOS-oriented aspects of science content courses.
(c) The use of argumentation in the classroom is one of the most novel aspects of research reviewed in this study. This research has centered on the historical context (e.g., controversies with respect to atomic models) and everyday life context (e.g., why does your skin get redder when you exercise) and analyzed within the philosophical frameworks of Toulmin (1958) and Lakatos (1970).

The previous study (Niaz 2012a, Chap. 3) showed that students' and teachers' understandings of the nature of science, in different cultures and continents, are quite similar to an empiricist epistemology. Similarly, it was found that the presentation of the atomic structure in science textbooks published in the USA, Venezuela, Turkey, and Korea is quite similar (Chap. 4, this book).

At this stage it would be interesting to explore the NOS views of science teachers in mainland China, for two reasons: (a) It does not form part of the traditional Western culture and it is almost a continent by itself. (b) Despite recent interchange with other countries, it has developed a culture, an educational system, and a very centralized government—with strong ideological commitments. Recently, science education researchers have established contacts with Chinese scholars and some studies have been published. A review of those studies is not possible here. However, I have selected one study (Wan et al. 2013) that can provide some information for comparing the NOS views of mainland Chinese science teacher educators and their counterparts in other countries.

Wan, Wong, and Zhan (2013) interviewed 24 science teacher educators from the economically developed areas of mainland China who had interest in talking about their conceptions of NOS. Some of these science teacher educators are also authors of school science textbooks and/or textbooks for training science teachers and have also participated in the development of the National Curriculum Standards in China. All the teachers participated in two face-to-face, semi-structured interviews, which started with the following open-ended question: How do you teach NOS to your prospective science teachers in your own course and why? This was followed by similar follow-up questions. Each interview lasted from 45 to 100 min. Table 3.2 presents a summary of the results (adapted from Wan et al. 2013, pp. 1126–1127). In this table I have included results of (a) NOS elements in which at least 50 % (12 out of 24) of the participants expressed this view (elements 1–5) and (b) NOS ele-

Table 3.2 NOS views of Chinese science educators (Adapted from Wan et al. 2013)

NOS elements suggested by Chinese science teacher educators	Frequency ($n = 24$)
1. Scientific investigation is based on observation and experiments. The validity of scientific claims aims to be settled by these empirical data	20
2. Scientific investigation relies on inductive and deductive logics to bridge between empirical data and scientific knowledge	15
3. Generally speaking, there exists a process of scientific investigation, like raising questions, hypothesizing, collecting data, analyzing data, drawing conclusion, and communicating.	12
4. The development of scientific knowledge is an accumulative and progressive process	14
5. Human inquiry into the nature is guided by realist beliefs of mind and nature, like the existence of an external world that is independent of the observer, the universality and constancy of connection in the world, and the possibility of our mind to know the external world and connections within it	13
6. Theory-laden nature of observation: Human observation cannot be absolutely objective. It is unavoidably influenced by the observer's theoretical and discipline commitments, beliefs, prior knowledge, training, experiences, and expectations	8
7. The development of scientific knowledge is a process of increasingly approaching to the truth	8
8. The development of science is influenced by complex social and cultural factors	10
9. Inferential nature of scientific knowledge	8
10. Science is being carried out in the cooperation among different scientists in modern society	9
11. The development of science requires perseverance, skepticism, objectivity, intellectual honesty, and selflessness	10

ments in which at least 33 % (8 out of 24) of the participants expressed this view (elements 6–11). These 11 NOS elements would perhaps reflect a significant part of the NOS views of Chinese science teacher educators. A comparison of these views with those of both students and science teachers in many other parts of the world shows a considerable amount of similarity (for a review see Niaz 2012a, Chap. 3).

In a Delphi style study, Osborne, Collins, Ratcliffe, Millar, and Duschl (2003) consulted 23 experts engaged in the study of science and its communication, based on the following groups: leading scientists; historians, philosophers, and sociologists of science; science educators; science teachers; and public understanding of science. These experts recommended that the following "ideas about science" could be taught in school science: (1) scientific method and critical testing, (2) creativity, (3) historical development of scientific knowledge, (4) science and questioning, (5) diversity of scientific thinking, (6) analysis and interpretation of data, (7) science and certainty, (8) hypothesis and prediction, and (9) cooperation and collaboration in the development of scientific knowledge. Interestingly, the experts assigned the highest priority to the teaching of "scientific method and critical testing," with reasons such as (a) "core process on which the whole edifice of science is built," (b) "central thrust of scientific research," and (c) "… careful experimentation is used to test hypotheses." In spite of some differences, especially with respect to the terminology used, Osborne et al. (2003) would recommend about the same set of "ideas about science" for inclusion in the science classroom, as many other science education researchers in different parts of the world. Nevertheless, there is an important and crucial difference with respect to what Osborne et al. (2003) refer to as "scientific method and critical testing" and, for example, Lederman et al. (2002) as "myth of the scientific method." It is important to note that most science curricula and textbooks in most parts of the world emphasize inductivism, falsificationism, and the scientific method. Given this perspective Osborne et al.'s (2003) emphasis on the scientific method seems to be an anachronism, as the experts in this study assigned the highest priority to this idea. Furthermore, the combining of "scientific method" and "critical testing" obscures the issue and even suggests that the use of the "scientific method" is perhaps inevitable. In contrast, Lederman et al. (2002) clearly traced its origin to Francis Bacon's *Novum Organum* and emphasize its unhealthy influence on school science (for details, see Niaz 2011, Chapter. 10).

At this stage, it is interesting to consider the following view of some (6 out of 24) Chinese science teacher educators, with respect to the scientific method: "There is no one method of science applicable at all times that can guarantee the development of infallible knowledge" (reproduced in Wan et al. 2013, p. 1126). This is an informed view of the scientific method that coincides with the views of many science education researchers (e.g., Jenkins 2007; Lederman et al. 2002; Windschitl 2004) and at the same time contrasts sharply with that of Osborne et al. (2003). Interestingly, Wan et al. (2013) compare (Table 1, p. 1118) the NOS elements recommended by Osborne et al. (2003) with those of McComas and Olson (1998), without mentioning the important finding of their own study in which six participants had an informed view of the scientific method that contrasts with that of Osborne et al. (2003). In order to understand further the views of Chinese science teacher educators with respect to the scientific method, let us examine a study reported by Windschitl (2004), in which 14 preservice secondary science teachers in the USA developed their own empirical investigations, from formulating questions to defending results in front of peers. Findings indicated that the teachers consistently used the following misrepresentations of some fundamental aspects of science: (a) A hypothesis functions as a guess about an outcome, but is not necessarily part of a larger explanatory system. (b) Background knowledge may be used to provide ideas about what to study, but this knowledge is not in the form of a theory or model. (c) A theory is an optional tool one might use at the end of a study to help explain results. Finally, Windschitl (2004) concluded that these ideas appear consistent with a "folk theory" of an atheoretical scientific method, "… *that is promoted subtly, but pervasively, in textbooks, through the media, and by members of the science education community themselves*" (p. 481, emphasis added). Indeed, a comparison of the three studies (Osborne et al. 2003; Wan et al. 2013; Windschitl 2004) conducted in three different countries (UK, mainland China, and USA) is thought-provoking, as it can help to further our understanding of NOS.

It appears that the NOS views of Chinese science teacher educators (see Table 3.2), and their counterparts in many other parts of the world, despite some differences, are quite similar (for a review see, Niaz 2012a, Chap. 3). How can we understand and explain this? A possible explanation is provided by Wan et al. (2013). In various parts of their article, they state that Marxism is the dominant philosophy in China, and most teachers and students have to take courses that deal with Marxist ideology and conclude categorically: "… empirical evidence, which is emphasized by empiricist epistemology, is implicitly integrated into the concept of *practice* in Marxism. Thus it is believed that Marxist philosophy, to a large extent, is consistent with empiricist epistemology" (Wan et al. 2013, p. 1120, italics in the original). Furthermore, for most practical purposes, Marxism is a Western philosophy. If these findings can be sustained in future research, then these have important educational implications, namely, that science educators in mainland China (despite their apparent isolation) face similar epistemological difficulties as their colleagues in many other parts of the world. Consequently, joint educational research programs can be mutually beneficial.

Teaching the Nature of Science in the Classroom

In this section I present examples of the following three studies conducted in different parts of the world that have included different elements of the nature of science in the chemistry classroom along with the historical context in which it is embedded: Bektas et al. (2013) in Turkey, Tolvanen et al. (2014) in Finland, and Niaz (2011) in Venezuela.

Bektas et al. (2013) designed a study to improve the NOS understanding of seven preservice chemistry teachers in Turkey. The purpose of the study was to facilitate teachers' understanding of two NOS elements (relationship between laws and theories and tentative nature of science), in the context of the particulate nature of matter. In Turkey, the curriculum does not specifically include NOS, and hence the authors decided to embed NOS elements in the chemistry course. Three weeks of the course (12 classroom hours out of the 14) were devoted to teaching NOS. During this period, students conducted activities related to the aspects of NOS such as "New Society" (Cavallo 2008) and "The Cube Activity" (Lederman and Abd-El-Khalick 1998). After the activities students participated in discussions and finally responded to open-ended questions followed by interviews. Results obtained showed improvement in students' understanding of NOS. For example, one student stated that the law of definite proportions was first stated by Proust, and then Dalton proposed his atomic model by using results of this law, which was considered an informed view of NOS with respect to laws and theories (for a historical background to this episode, see Chap. 4). Another student suggested a teaching approach based on emphasizing the differences in the thought processes of Aristotle, Democritus and Dalton, Thomson, Rutherford, and Bohr, which shows how theories are not necessarily "true" and can change continuously. This is an interesting response as it refers not only to the various atomic models/concepts but also brings up the issue of how we need to understand the difference between, say the Greek philosophers and the modern theorists. This issue is currently the subject of debate among historians and philosophers of science (Chalmers et al. see Chap. 4 for a discussion and evaluation of this topic in general chemistry textbooks).

Another student referred to the tentative nature of science by responding:

> There are no absolute facts in science. For example: as we know, there are many atomic models proposed in science … After Dalton proposed his atom model, other scientists continue[d] to work … and proposed different models. If there were absolute facts, scientists would accept Dalton's truth as absolute … model. (Bektas et al. 2013, p. 206)

This is a novel way to approach the issue of the tentative nature of science. For example, after discussing the various atomic models in the classroom, a teacher could set the stage for an interesting discussion by asking: If there is an unchanging absolute view of understanding the particulate nature of matter, why did Dalton's and the subsequent models change continuously? Based on their experience of teaching particulate nature of matter to Turkish preservice chemistry teachers, Bektas et al. (2013) have recommended that teachers be given opportunities to integrate NOS into their teaching practice and, furthermore, NOS be embedded in specific science content.

Tolvanen et al. (2014) have analyzed curriculum materials based on the history of science. In order to use historical materials in chemistry education, they asked the following research questions: (1) What NOS content is included in the lesson plans? (2) How historical experiments are used to teach NOS? (3) How historical narratives are used in the lesson plans? The study is based on the participation of 16 secondary school chemistry teachers in Finland, as part of their master's degree program in chemistry education. Lesson plans were elaborated based on a review of the historical literature, and the following NOS elements along with the historical context were discussed:

1. *Tentative*: To show that scientific knowledge is durable but uncertain, the work of Lavoisier was discussed to show that some of his ideas were correct and others were later proven wrong.
2. *Difference between laws and theories*: Mendeleev's periodic law was used as an example, and his reaction to the discovery of noble gases and radioactivity, both of which went against Mendeleev's theories, was described. (Note: Authors refer to Mendeleev's work first as a law and later as a theory. There is some controversy in the literature with respect to Mendeleev's contribution: theory or a law. For details see Niaz et al. 2004 and also a section in Chap. 7 with respect to the periodic table.)
3. *Empirical*: The experimental work of Proust and Berthollet and their relationship to the development and acceptance of Dalton's atomic theory were presented. (Note: For further details related to Dalton's theory, see Niaz 2001b and also Chap. 4.)

4. *Model-based*: Pros and cons of Bohr's and Lewis' atomic models were presented, and it was discussed whether there is one right way to present models, with no reference to rivalry or controversy.
5. *Inferential*: Theories as tools to interpret concepts, such as "heat," "energy," and "falling bodies."
6. *Creativity*: Avogadro's hypothesis as an example of how creative insight is crucial in science. Furthermore, students were asked to discuss how Mendeleev's work illustrates that doing science requires creativity and imagination.
7. *Social dimensions*: (a) collaboration among scientists; role of the Royal Society in Boyle's work and the Karlsruhe Conference showed that science is not a solitary undertaking; friendship between Bunsen and Kirchoff; (b) power and politics within the scientific community; slow acceptance of Avogadro's hypothesis by the scientific community, Berzelius as an example of an influential scientist who hindered the acceptance of Avogadro's rival hypothesis.
8. *Societal dimensions*: (a) influence of the larger cultural milieu in scientific practice and knowledge and how chemical reactions were written differently in earlier periods of time; (b) relationship between science and industry and how the requirements of agriculture and industry guided Fritz Haber's research on ammonia synthesis; and (c) the relationship between chemistry and commerce and the economic impact of indigo and dye-producing countries.
9. *Instrumentation*: Role of scientific instruments in the empirical nature of chemistry; how the Bunsen burner played a crucial role in the development of spectroscopy.

It is important to note that the nine NOS elements used by Tolvanen et al. (2014) to introduce the historical approach in the chemistry classroom are quite similar to those of Lederman et al. (2002) and McComas and Olson (1998). Furthermore, there are some differences, such as the separation of "social and historical milieu" into two parts, namely, social and societal dimensions. However, the most important innovation of Tolvanen et al. (2014) is the inclusion of domain-specific aspects related to each of the nine NOS elements, which are based on the chemistry curriculum. Classroom discussions based on such domain-specific issues can be much more helpful for students to understand progress in chemistry, rather than simply enumerating certain domain-general NOS elements. A word of caution is, however, necessary as one could disagree with some details of the domain-specific aspects, such as whether Mendeleev's contribution can be considered as a theory or a law (this debate is continued Chap. 7). This also illustrates that just like history of science, history of chemistry is complex and is in a continuous state of controversy and evolution. This book provides many examples of such issues that are currently being debated. For example, consider the views of Chalmers, Needham, and Rocke with respect to Dalton's atomic theory discussed in Chap. 4. This clearly shows that if we decide to wait for the historians and philosophers of science to agree and reach a consensus, our classroom practice would perhaps never change.

Finally, Tolvanen et al. (2014), based on their research experience of elaborating science lessons in collaboration with chemistry teachers, consider that the following guidelines can be helpful in introducing the historical approach:

1. Explicit discussion of specific NOS issues is preferable to a general discussion of multiple NOS aspects.
2. Comprehensive use of narratives is recommended. All parts of the lesson (e.g., exercises, laboratory work) should be connected to the narrative, historical experiments and the curriculum material.
3. Clear instructions on how to use the narrative in discussing the NOS issues with the students.
4. In order to enhance the narrative effect of the historical account, it is desirable to include at least one conflict or controversial aspect.

Niaz (2011, Chap. 9) has designed a study to facilitate in-service chemistry teachers' understanding of the nature of science based on historical controversies. A basic premise of the study was that a discussion of chemistry content within the historical context could help teachers to discuss, argue for or against a particular interpretation of experimental evidence, and finally deepen their understanding of various aspects of the nature of science. This study is based on 17 in-service chemistry teachers (secondary and introductory university level), who had enrolled in a course as part of their master's degree program in chemistry education at a major university in Venezuela. Among other readings based on history and philosophy of science, the following were of particular interest for the participants: (a) Niaz (1998b), which deals with the historical reconstruction of the atomic models of Thomson, Rutherford, and Bohr and an evaluation of these models in general chemistry textbooks (of particular interest in this reading were the controversies that ensued following the presentation of these atomic models, e.g., Thomson–Rutherford and Rutherford–Bohr), and (b) Niaz (2000a), which deals with the historical reconstruction of the determination of the elementary electrical charge based on the oil drop experiment (of particular interest was the controversy between R. Millikan and F. Ehrenhaft). During the course (11 weeks), participants made presentations, participated in discussions, and responded to written questions. After this experience (classroom discussions), the following NOS aspects emerged:

(a) *Scientific method*: For some participants at the beginning of the course, this was a simple and straightforward way to understand how science was done. This conceptualization slowly started to change and most participants at the end of the course realized that this was a "caricature" of what real science is. The following are three examples of participants' responses that were considered to be informed views of the scientific method (reproduced from Niaz 2011, p. 142):

> In view of the universality and rigidity of the scientific method, one could believe that: "Science does not change." For some it may signify that if science changes, *it does not exist*. (Emphasis in the original)

> … some textbook authors postulate the scientific method not as an alternative but rather as obligatory for the scientist …

> Chemistry needs to be "freed" of myths and history and philosophy of science could help. It needs to be emphasized that there is no one scientific method, but rather diverse methods and processes—textbooks cannot continue to be a list of questionnaires and algorithmic problems and answers.

These three responses indicate different degrees of a critical understanding of the scientific method. The first response shows concern with respect to how a rigid view of the scientific method may lead us to believe that if science changes then it may not exist. The second example focuses attention on how the textbooks convey the message that scientists always follow the scientific method. The third example considers the need for freeing chemistry from a "myth" and invokes the perspective provided by the history and philosophy of science (cf. Lederman and Abd-El-Khalick 1998).

(b) *Empirical nature of chemistry*: Again this was an important issue during class discussions, as most textbooks and their previous chemistry courses had emphasized this aspect of chemistry very thoroughly. The following are two examples of how the participants viewed this issue after the course (reproduced from Niaz 2011, p. 142):

> Chemistry is considered to be an experimental science in which laboratory work leads to the production of knowledge with no reference to controversies and debates that help to construct scientific knowledge.

> No effort is made to differentiate between the idealized scientific law and the observations—as a consequence students tend to memorize the laws.

Both responses clearly show how the participants' understanding went beyond the traditional chemistry textbooks and courses. To emphasize the experimental nature of chemistry is correct. However, to ignore the controversies and debates in chemistry (as most textbooks and course materials do) comes quite close to "distorting" the history of chemistry. Most philosophers of science would agree that scientific laws do not refer to actual bodies or phenomena but rather to an idealization that can be approximated in the laboratory. However, the most important part of this response is that it establishes a relationship between this lack of an epistemological perspective and the memorization of laws by the students. Interestingly, Stephen Brush (1978) has referred to this aspect of chemistry in cogent terms:

> Of course, as soon as you start to look at how chemical theories developed and how they were related to experiments, you discover that the conventional wisdom about the empirical nature of chemistry is wrong. The history of chemistry cannot be used to indoctrinate students in Baconian methods (p. 290).

(c) *Objectivity in science*: This aspect of NOS was clearly complex and, at the same time, very interesting. Participants were provided the following question/dilemma:

> Martin Perl, Nobel laureate in physics 1995, in his search for the fundamental particle (quark) has elaborated a philosophy of speculative experiments: "Choices in the design of speculative experiments usually cannot be made simply on the basis of pure reason. The experimenter usually has to base her or his decision partly on what feels right, partly on

what technology they like, and partly on what aspects of the speculations they like" (Perl and Lee 1997, p. 699). Given the methodologies of Thomson, Rutherford, Bohr, Millikan and Ehrenhaft (Niaz 1998b, 2000a), in your opinion, what are the implications of this statement for teaching chemistry? (Reproduced in Niaz 2011, p. 132)

The rationale behind using this episode from the history of science was to present an experience from a current leading scientist working on cutting-edge experimental work and how a scientist goes about coping with difficulties (for details with respect to Perl's methodology, see Niaz 2012a, Chap. 7). The reference to Perl's experimental methodology is important as some students may think that what scientists did in the past (e.g., Thomson, Rutherford, Bohr, Millikan) was perhaps very different from what scientists do these days. The following are three examples of participants' responses to this question (Reproduced in Niaz 2011, pp. 139–140):

> According to Lakatos, theories can "live" together for some time and after a period of arguments and confrontation the scientific community decides in favor of one or the other. Similarly, it is probable that Martin Perl considers the conjugation of speculation and reason as an important element in looking for an answer to a particular question. In the Millikan-Ehrenhaft controversy, Millikan based on the "negative heuristic" of his research program decided to discard some of the data. This was perhaps a recognition, that besides reason, speculation and intuition also played an important part… A similar process occurred in the case of the atomic theories [Thomson, Rutherford, Bohr] … This shows that everything cannot be solved by logic, and it is necessary to look for other alternatives provided they are consistent and well justified … Far from confusing the students, these episodes can arouse their curiosity and hence interest in science.

> … statement by Perl helps to "humanize" chemistry … it opens a new window with respect to scientific knowledge … discussion of such issues in the classroom can facilitate conceptual change towards constructivist views … it will also require innovative teaching strategies …

> … Millikan did not manifest in public the speculative part of his research … Perl, however has affirmed publicly that at times he speculates … Perl's affirmation manifests what Millikan in some sense tried to "conceal", viz., science does not develop by appealing to objectivity in an absolute sense and that science does not have an explanation for everything and hence the need for research. Acceptance of the fact that science does not have an absolute truth and nor an immediate explanation for everything, would change students' conception of science and chemistry in particular. This will show chemistry to be a science in constant progress and that what is true today may be false tomorrow and may even help to originate a new truth—sequences of heuristic principles. [cf. Burbules and Linn 1991]

An important aspect of all these responses is that the participants did not simply reiterate what Perl had stated but rather tried to understand the dilemma within a particular context of the chemistry curriculum and their own experience. The following are some examples: (i) Despite arguments and confrontations, theories can "live" together for some time (note that this is attributed to Lakatos but the wording is that of the participant and hence the originality and the creative effort). (ii) Perl's conjugation of speculation and reason is compared to how two theories can "live" together. The concept of conjugation is not invoked by Perl, but is an innovation of the participant. (iii) The fact that Millikan discarded some of his data is attributed to his "negative heuristic" (cf. Lakatos 1970). (iv) Millikan's methodology is compared to that of speculation and intuition, similar to Perl. (v) Perl's statement can help to "humanize" chemistry and could help teachers to innovate with respect to new teaching strategies. (vi) Millikan did not manifest in public the speculative part of his research and may even have tried to "conceal" (cf. Holton 1978a for the Millikan–Ehrenhaft controversy). (vii) These episodes show that "objectivity in science" will have to be understood in a particular context of scientific progress. In this context Machamer and Wolters (2004) advice is particularly helpful: "… to save the objectivity of science, we must free it from an ideal of rationality modeled after mathematics and logic …" (p. 9).

Next Generation Science Standards (NGSS) and Nature of Science

Elaboration of the Standards has gone through a long process of peer and institutional evaluation and most scholars would agree with the opening statement with respect to the nature of science: "Scientists and science teachers agree that science is a way of explaining the natural world. In common parlance, science is both a set of practices and the historical accumulation of knowledge … Further, students should develop an understanding of the enterprise of science as a whole—the wondering, investigating, questioning, data collecting and analyzing" (NRC 2013, NGSS, Appendix H, p. 1). Most science educators and researchers would perhaps consider this to be a consensus statement.

Now let us consider the NOS matrix (p. 4) with respect to the basic understandings about the nature of science:

1. Scientific investigations use a variety of methods.
2. Scientific knowledge is based on empirical evidence.
3. Scientific knowledge is open to revision in light of new evidence.
4. Scientific models, laws, mechanisms, and theories explain natural phenomena.
5. Science is a way of knowing.
6. Scientific knowledge assumes an order and consistency in natural systems.
7. Science is a human endeavor.
8. Science addresses questions about the natural and material world.

The first four refer to nature of science understandings closely associated with practices, and the last four refer to understandings most closely associated with crosscutting concepts. Tables on pages 5 and 6 (Appendix H) then provide learning outcomes for the different grade bands. Of particular interest for this book were those pertaining to the high school band (grade levels). Most of these learning outcomes are carefully crafted, and perhaps most science education researchers would agree with them. However, in these tables (pages 5 and 6), among others, there is no explicit reference to the following important NOS aspects: (a) scientific method; (b) role of rival theories which explain the same observations leading to competition, conflicts, and controversies (e.g., interpretation of alpha-particle experiments by J.J. Thomson and E. Rutherford); and (c) underdetermination of scientific theories by experimental evidence, namely, no amount of experimental evidence can provide conclusive proof for a theory (e.g., standard quantum mechanics and Bohmian mechanics). These are controversial issues and precisely for this reason need our attention. For example, most science textbooks present the scientific method as a "panacea," and teachers generally endorse it in their classroom activities. Actually, even science education researchers differ on its importance in the science curriculum. According to Osborne et al. (2003, p. 706), the scientific method constitutes the "central thrust of scientific research." In contrast, Lederman et al. (2002, p. 501) consider the "recipe-like stepwise procedure" as a myth. Given that Appendix H includes both articles in its reference list, some guidance on this issue would have been helpful for science teachers. In a study (Niaz 2011, Chap. 10) with 17 in-service chemistry teachers (both high school and introductory university level), some of them expressed their concern and suggested that these two groups of science education researchers could help to clarify this apparent contradiction in their understanding of the scientific method.

Despite these considerations, I found the following statement interesting and quite close to the recognition of conflicts, controversies, and contradictions: "Respectfully provide and/or receive critiques on scientific arguments by probing reasoning and evidence and challenging ideas and conclusions, responding thoughtfully to diverse perspectives, and determining what additional information is required to resolve contradictions" (NGSS 2013, Appendix F, p. 29). Nevertheless, I was intrigued by the inclusion of the word "respectfully" in this statement, as criticisms should normally be respectful of the alternative point of view. Overall, NGSS has included some novel ideas, such as performance expectations, learning progressions, and practices (instead of skills, which approximates more to scientific practice), and hopefully further improvements will be made, based on feedback from teachers and researchers.

Since the release of the final version of NGSS (NRC 2013) in April 2013, it has been the subject of analyses by the US-based National Association for Research in Science Teaching (NARST) (www.narst.org). Being an international organization, a major strength that NARST brings to this challenge is that its members can provide the necessary research experience to enhance the implementation of the NGSS. Position papers prepared by eight research teams have recently been released dealing with accountability, assessment, curriculum materials, engineering, equity, informal science education, preservice science teacher education, and professional development (for details, see Lynch and Bryan 2014). Despite the generally positive reception of NGSS, a lot of work still remains to be done, especially with respect to teacher training. This issue has been raised by Lederman and Lederman (2013) in cogent terms:

> The NGSS is truly an ambitious vision for K-12 science education. But, as has been the case with previous calls for change, there is little conversation about the knowledge and ability of science teacher educators to help facilitate the change. Science teacher education, as well as teacher education in general, is consistently under attack by policy makers and other

> stakeholders. If we are truly a part of the solution to improve the quality of teaching and learning of science, we must carefully consider the specific elements of professional development that is needed for our current and future science teacher educators, and the qualifications and abilities of those who will deliver the professional development (p. 931).

The importance of NGSS is the subject of discussions at international meetings. For science education, it is particularly important for understanding the nature of science and teacher training, and Lederman and Lederman (2013) have clearly emphasized this aspect in NGSS.

Chapter 4
Understanding Atomic Models in Chemistry: Why Do Models Change?

Introduction

The importance of understanding atomic structure is recognized for all high school and introductory university-level chemistry courses in almost every part of the world. Despite research in science education, history, and philosophy of science, it continues to be a difficult and controversial topic. Justi and Gilbert (2000) analyzed high school chemistry textbooks (nine from Brazil and three from the UK; published during 1993–1997) to study the presentation of atomic models. These authors reported the use of hybrid models in textbooks based on various historical developments, such as ancient Greek, Dalton, Thomson, Rutherford, Bohr, and quantum mechanics (Schrödinger's equation). Hybrid models do not provide students with an opportunity to understand the dynamic nature of science, in which different approaches to understand phenomena are contrasted and critiqued. The authors concluded: "Hybrid models, by their nature as composites drawn from several distinct historical models, do not allow the history and philosophy of science to make a full contribution to science education" (p. 993).

The nub of the topic revolves around the following question: Why do atomic models change? The authors of a general chemistry textbook expressed this succinctly: "The story of the development of the modern model of the atom is an excellent illustration of how models are *constructed and revised*" (Atkins and Jones 2002, p. F15, emphasis added). The idea of *construction* and *revision*, of course, has philosophical overtones and will be the subject of this chapter. In a similar vein, in order to understand the importance of atoms, one general chemistry textbook reproduced the following quote from Nobel Laureate Richard Feynman (1918–1988): "The most important hypothesis in all of biology is that everything that animals do, atoms do. In other words, there is nothing that living things do that cannot be understood from the point of view that they are made of atoms acting according to the laws of physics" (p. 51, reproduced in Tro 2008).

M. Niaz, *Chemistry Education and Contributions from History and Philosophy of Science*, Science: Philosophy, History and Education,
DOI 10.1007/978-3-319-26248-2_4

Origin of Atomic Theory (From Democritus to Dalton)

The role played by ancient Greek philosophy (e.g., Leucippus and Democritus) has been the subject of research and controversy in the science education literature. Sakkopoulos and Vitoratos (1996) have explored the empirical foundations of atomism in ancient Greek philosophy by considering the following example from Democritus:

> Cone or cylinder? Democritus put the following question: "How must we imagine the two circular surfaces resulting from the section of a cone by a plane parallel to its base?" If the answer is that the two circles are equal, then the cone must turn into a cylinder. To prevent this, one had to accept that the two circles must be unequal. The lowest limit cannot be zero and this leads to the atomic hypothesis. The minimum difference must be equal to an atom. (p. 301)

According to these authors, the concept of atom was a great invention of the human mind. Despite the lack of experimental evidence, it provided explanations of various everyday phenomena, such as the orderly growth and decay of humans, animals, and plants, the spreading of a scent, the evaporation and condensation of water, and the wearing out of a pavement by the steps of passersby. All these examples, if discussed in class, can show that an idea can provide an explanation, which at best is tentative and would eventually be replaced by a better explanation. Chalmers (1998), however, has disputed this claim and considered the Greek atomic theory as an obstacle to, rather than an anticipation of, modern science. On the contrary, Irwin (2000) considers the Greek ideas about the atom as an opportunity to understand the origin of an idea and the role that it can play to familiarize students with the development and progress of scientific theories. Furthermore, the Greek methodology of finding support for the atoms (e.g., cone or cylinder) can be considered as a series of thought experiments. As an example of thought experimentation, Irwin asked students to imagine an iron nail being cut in half, followed by one of the halves being cut in half, and so on, until no further division was possible. At the end, the following question was posed: Is there a point where we reach a particle that is indivisible? (Irwin 2000, p. 15)

More recently, Chalmers (2009) has been even more critical of ancient Greek philosophy and called for clearly differentiating between the atoms of the ancient Greeks, Boyle's atoms, Dalton's atomic theory, and modern atomic theory:

> The atoms invoked by Ancient Greeks such as Democritus and Epicurus and by seventeenth-century mechanical philosophers such as Gassendi and Boyle were construed as the ultimate and unchanging components of material reality, … Twentieth-century atoms are nothing like those *envisaged* in these philosophical traditions and they and their properties were discovered by experiment rather than philosophical analysis. The modern atom has an internal structure, most importantly an electron structure … Such properties are far from anything *envisaged* by Democritus and Boyle …. (p. 262, italics added)

Chalmers' main argument seems to be that the modern atom is very different from what Democritus and Boyle could have *envisaged*. Despite its apparent appeal, this leads to various paradoxical issues, such as: a) Could J. J. Thomson have *envisaged* in 1897, the electronic structure as it came to be established? b) Could E. Rutherford (and Thomson, too) have *envisaged* the quantum mechanical atom in 1911? c) Could N. Bohr have *envisaged* the wave-mechanical atom in 1913? Despite these difficulties with respect to *envisaging* an atomic model yet to be developed, most historians, philosophers of science, chemists, textbook authors, and students would recognize the contributions of Thomson, Rutherford, Bohr, and others toward the postulation of the modern atomic theory. Consequently, if we do recognize the merits of Thomson, Rutherford, and Bohr, why should we deny the same to the ancient Greek philosophers? Nevertheless, it is important to recognize the difference between the atomic models of the ancient Greeks (Democritus), Boyle (seventeenth century), Dalton (nineteenth century), and the modern atomic theories. From the pedagogical point of view, the transition from one model to another is an important aspect of progress in science and facilitates an understanding of the nature of science.

In this context, it is important to note that Viana and Porto (2010) have suggested that we need to go beyond the historiographical approach prevalent in the first half of the twentieth century, in which students were presented naïve versions of Dalton's theoretical formulations based on inductivism. Starting in the 1960s, it was the historicist school that facilitated a better understanding of the complexities involved in scientific progress and the nature of science.

Dalton as a Dilemma

In this section, I will contrast the views of two eminent scholars: A. Chalmers (philosopher of science) and A. Rocke (a historian). I sent the following email (October 30, 2013) to Rocke:

> I am writing to seek your opinion with respect to the following statement from Chalmers (2009): "… there is much that is mistaken and misleading about seeing Dalton's atomic theory as the beginnings of an experimentally testable version of atomism … Dalton's theory had no testable content that went beyond the laws of proportion that it entailed and so could not productively guide chemistry in a way that could not be achieved by way of the laws of proportion alone" (pp. 173–174). In my opinion, this contrasts with your statement: "One of my central concerns here has been to explode the persistent myth, as prevalent today as it was in the nineteenth century, that there existed a nonatomic chemistry which formed a viable alternative to the Daltonian system." (Rocke 1978, p. 262)

Rocke (2013a) responded within 3 h in the following terms:

> It will not surprise you when I tell you that I think that Chalmers (and Needham) are wrong. In my opinion it is crucial to pay close attention not just to the laws of constant and multiple proportions, but also to the law of equivalent proportions. In fact, I argue that the laws of constant and multiple proportions are really nothing more than special cases of the law of equivalent proportions, so that really there is only this one law of stoichiometry. But this one law is almost ignored in these discussions. I do believe that the laws of stoichiometry are proper laws, and that Dalton's atomism is not only a theory, but a successful theory at that [Needham refers to the eminent philosopher of science Paul Needham].

Needham (2004a, b, 2010) has argued cogently that Daltonian atomism has not provided chemistry with any explanations. Similarly, according to Chalmers (2009, 2010), the progress made by chemistry in the nineteenth century owed little to Daltonian atomism. In contrast, Rocke (2013b) has claimed just the opposite: "I want to offer here a contrary view: I propose to rescue nineteenth-century atomic theory from the charge of irrelevance or even meaninglessness. I claim that atomic theory was, from the beginning, not only a robust and heuristically powerful theory, but crucial to the spectacular development of chemistry in that century" (p. 146).

In order to accomplish this task, Rocke has retranslated and reinterpreted the work of the German theoretical chemist H. Kopp (Rocke and Kopp 2012). Kopp spent a large part of his career, from 1843 to 1890, teaching chemical theory and published his book *Theoretische Chemie* in 1863. According to Kopp, there existed three kinds of theories in science:

(a) Those exemplified by Newtonian mechanics and kinetic–molecular theory, which posited certain principles from which a range of real phenomena could successfully be deduced. Kopp conceded that chemistry as yet had no such theory.
(b) Those that provided a recapitulation of experimental data under a mathematical generalization or rule, exemplified by the work of Dulong–Petit and Gay-Lussac's law of combining volumes.
(c) Picture of nature, viz., how one can think about phenomena.

According to Kopp, chemistry could lay claim to the second and third type of theories, and Dalton's atomic theory would be one example of such a theory. Furthermore, like any emerging theory, Dalton's contemporaries found flaws in its formulation. The chemical literature of the nineteenth century showed considerable engagement with such theories, and the work of Klein (2003) has demonstrated the heuristic power of early atomic theories in the progress of organic chemistry.

Returning specifically to Dalton, Rocke (2013c) has critiqued both Needham and Chalmers for having overemphasized the importance of the laws of definite and multiple proportions and for having ignored the law of equivalent proportions, which is crucial to all theories of chemical atomism:

> Actually, in a strict sense there is only one, not three laws of stoichiometry, for the law of definite proportions and the law of multiple proportions can be considered as special cases of the law of *equivalent* proportions … [which] leads so naturally to the idea of chemical atoms that nineteenth-century chemists, even the most anti-theoretical among them, could hardly help themselves. (p. 148)

This interpretation of the laws of stoichiometry will be helpful in Chap. 5. In this context, the following statement from Chalmers (2009) acknowledges the merits of Rocke's recognition of chemical atomism:

> I have most difficulty in defending my position when confronted by historians such as Rocke and Klein who see the nineteenth-century advances as coming about by way of a chemical atomism, rather than physical atomism in the tradition of Newton or the mechanical philosophers. (p. 264)

Finally, this section shows that all three (Chalmers, Needham, and Rocke) do recognize the merits of each other's respective philosophical positions.

Presentation of Dalton's Atomic Theory in General Chemistry Textbooks

Considering the historical reconstruction presented in the previous section, the objective of this section is to present an analysis of general chemistry textbooks, published in the USA, based on the following criteria (this work has not been published previously):

Criterion D1: Dalton Versus the Greek Philosophers

This criterion contrasts the methods of Dalton with respect to that of the ancient Greeks (especially of Democritus). It is important to understand that Dalton had access to some experimental evidence, whereas Democritus and other Greek philosophers based their concept of the atom on everyday observations, ideas, and thought experiments. A presentation of this topic in a textbook was classified as *Satisfactory* (S) if it explicitly described the difference between the methods of Dalton and the Greeks. This distinction is important if we want students to understand that both the methods and the atomic models can change over time. Textbooks that simply mentioned either Dalton's model of the atom or that of the Greeks were classified as *Mention* (M). Textbooks that did not mention either of the two methods or the models were classified as *No mention* (N). Table 4.1 presents the results based on the analyses of 32 general chemistry textbooks published (between 1999 and 2014) in the USA. Of the 32 textbooks analyzed, 20 (63 %) textbooks were classified as Satisfactory, 4 as Mention, and 8 as No mention. The following are examples of textbooks that were classified as Satisfactory:

> The Greeks asked what would happen if they continued to cut matter into ever smaller pieces. Is there a point at which they would have to stop because the pieces no longer had the same properties as the whole or could they go on cutting forever? We now know that there is a point at which we have to stop. That is, matter consists of almost unimaginably tiny particles. The smallest particle of an element that can exist is called an **atom** … The first convincing argument for atoms was made in 1807 by the English school teacher and chemist John Dalton … He made many measurements of the ratios of the masses of elements that combine together to form the substances we call 'compounds' … and found that the ratios formed patterns. (Atkins and Jones 2008, F16, emphasis in the original)

> Today our ideas are based on evidence. Democritus had no evidence to prove that matter cannot be divided an infinite number of times, just as Zeno had no evidence to support his claim that matter can be divided infinitely. Both claims were based not on evidence but on visionary belief: one in unity and the other in diversity … In 1808 the English chemist John Dalton (1766–1844) put forth a model of matter that underlies modern scientific atomic theory. The major difference between Dalton's theory and that of Democritus … is that Dalton based his theory on evidence rather than on a belief. (Bettelheim et al. 2012, pp. 34–35)

Table 4.1 Evaluation of Dalton's atomic theory in general chemistry textbooks based on a history and philosophy of science framework ($n=32$)

		Criteria[a]			
No.	Textbook	D1	D2	D3	Points[b]
1.	Atkins and Jones (2002)	S	N	N	2
2.	Atkins and Jones (2008)	S	N	I	3
3.	Bettelheim et al. (2012)	S	N	N	2
4.	Bishop (2002)	N	N	N	0
5.	Brady et al. (2000)	S	L	I	5
6.	Brown et al. (2012)	S	L	N	4
7.	Chang (2010)	S	L	I	5
8.	Denniston et al. (2011)	M	N	N	1
9.	Dickson (2000)	S	N	I	3
10.	Ebbing and Gammon (2012)	M	L	I	4
11.	Frost et al. (2011)	N	N	N	0
12.	Goldberg (2001)	S	L	I	5
13.	Hill and Petrucci (1999)	M	L	I	4
14.	Jones and Atkins (2000)	S	N	I	3
15.	Malone (2001)	S	N	N	2
16.	McMurry et al. (2007)	N	N	N	0
17.	McMurry and Fay (2001)	S	L	N	4
18.	McQuarrie et al. (2011)	N	I	I	2
19.	Moore et al. (2002)	N	L	I	3
20.	Moore et al. (2011)	N	L	I	3
21.	Oxtoby et al. (2012)	S	L	I	5
22.	Raymond (2010)	S	L	N	4
23.	Russo and Silver (2002)	S	N	N	2
24.	Seager and Slabaugh (2013)	M	N	N	1
25.	Silberberg (2000)	S	I	I	4
26.	Spencer et al. (2012)	N	N	I	1
27.	Stoker (2010)	N	N	N	0
28.	Timberlake (2010)	S	N	N	2
29.	Tro (2008)	S	L	N	4
30.	Umland and Bellama (1999)	S	I	I	4
31.	Whitten et al. (2013)	S	I	N	3
32.	Zumdahl and Zumdahl (2014)	S	L	I	5

Note: references to textbooks in this table are presented in Appendix 1

[a]Criteria: D1 (Dalton versus the Greek philosophers), D2 (Dalton and the law of multiple proportions), D3 (Dalton and Gay-Lussac's law of combining volumes)

[b]Points: Satisfactory (S) = 2 points, Lakatosian (L) = 2 points, Mention (M) = 1 point, Inductivist (I) = 1 point, No mention (N) = 0 point

> The concept of atoms began nearly 2500 years ago when certain Greek philosophers expressed the belief that matter is ultimately composed of tiny indivisible particles, and it is from the Greek word *atomos*, meaning 'not cut,' that the word atom is derived. The philosophers' conclusions, however, were not supported by any evidence; they were derived simply from philosophical reasoning … At the beginning of the nineteenth century, John Dalton (1766–1844), an English scientist, used the Greek concept of atoms to make sense out of the laws of conservation of mass and definite proportions. Dalton reasoned that if atoms really exist, they must have certain properties to account for these laws. (Brady et al. 2000, pp. 46–48)

> Democritus asked a hypothetical question: What happens if a sample of matter is divided into smaller and smaller bits? Would such subdivision reach an ultimate particle, or could the subdivision occur indefinitely and still be characteristic of the sample? Democritus favored the particle view and developed an atomic vision of matter as part of his philosophy … The philosophical idea of atoms was not based on reproducible experimental evidence and measurements … In 1803, John Dalton, drawing from the work of many early scientists, proposed a theory or model of the particulate nature of matter. (Dickson 2000, pp. 55–66)

> The theory of the atom has had a long history. The ancient Greeks postulated that matter exists in the form of atoms, but they did not base their theory on experiments, nor did they use it to develop additional ideas about atoms. In 1803, John Dalton proposed the first modern theory of the atom, which was based on the experimentally determined laws of conservation of mass, definite proportions, and multiple proportions. (Goldberg 2001, p. 78)

> Without the ability to gather data and do experiments, you can't employ the scientific method. For this reason, Democritus was unable to support his original atomic theory. In contrast, some of the most beautiful examples of the application of the scientific method . and the replacement of old theories with new ones come from the development of modern atomic theory, beginning with John Dalton in the early 1800s. When you read about this development in Chaps. 3 and 4, keep the scientific method in mind, and you will understand why theories came and went. (Russo and Silver 2002, p. 18)

> The Greek philosopher Democritus (470–400 BC) suggested that all matter is composed of tiny, discrete, indivisible particles called atoms. His ideas, based entirely on speculation rather than experimental evidence, were rejected for 2000 years. By the late 1700s, scientists began to realize that the concept of atoms provided an explanation for many experimental observations about the nature of matter … Dalton's explanation summarized and expanded the nebulous concepts of early philosophers and scientists; more importantly, his ideas were based on *reproducible experimental results* of measurements by many scientists. These ideas were the core of *Dalton's atomic theory*, one of the highlights in the history of scientific thought. (Whitten et al. 2013, pp. 5–6, italics in the original)

These presentations from general chemistry textbooks constitute an interesting mosaic of how the historical context can be approached from different perspectives. Despite the different approaches, all these textbooks highlight the original ideas of Greek philosophers and contrast them with those of Dalton. This contrasts with Chalmers's (1998, 2009) thesis that Greek atomic theory was an obstacle to, rather than an anticipation of, modern science. Actually, the degree to which textbooks include the ideas of ancient Greek philosophers and then compare them with those of Dalton is quite surprising and at the same time provides a glimpse of the origin of ideas and their development. Some of the salient aspects of these presentations that provide a rationale for Dalton's atomic theory are the following:

(a) According to the ancient Greek philosophers, if we cut matter into smaller pieces, there is a point at which we have to stop, and hence Dalton's theory makes sense.
(b) Dalton based his theory on evidence, whereas that of Democritus was based on visionary belief.
(c) Based on the philosophical arguments of the Greeks, Dalton reasoned that if atoms really exist, they must have certain properties to account for the laws of conservation of mass and definite proportions.
(d) Based on hypothetical questions, Democritus developed an atomic vision of matter, for which he did not have reproducible experimental evidence. It was such ideas that helped Dalton to postulate his theory of the particulate nature of matter,
(e) The development of atomic theory shows how the scientific method helped the replacement (Democritus to Dalton) of old theories with new ones.
(f) Ideas of Democritus were based on speculations, whereas Dalton explained these nebulous concepts based on reproducible experimental results.

These aspects reveal that the authors of most general chemistry textbook do understand that the ideas of the ancient Greek philosophers (particularly Democritus) were based on visionary beliefs, philosophical arguments, hypothetical questions, and speculations. Furthermore, they clearly differentiate between the methodology of the Greeks and the reproducible experimental evidence used by Dalton. This clearly shows that these textbooks did not consider the philosophical ideas of the ancient Greeks to be an obstacle in the development of the atomic theory but rather invoked these ideas to understand Dalton's atomic theory (cf. Chalmers 1998, 2009; Irwin 2000; Sakkopoulos and Vitoratos 1996.

Criterion D2: Dalton and the Law of Multiple Proportions

Some scholars (such as Thomson 1825) in the early nineteenth century popularized the positivist version that Dalton was led to his atomic theory by the discovery of the law of multiple proportions while working on two hydrocarbons (methane and ethane). According to Rocke (1984), "This inductivist version was quite concordant with the then prevalent Victorian model of heroic science" (p. 27). Linus Pauling (1964) has clarified the issue by stating categorically that the law of simple multiple proportions was derived from the theory and then tested by experiments. The objective of this criterion is to evaluate if textbooks follow one of the following interpretations with respect to the law of multiple proportions:

(a) *Inductivist* (I): Dalton was led to his atomic theory by the discovery of the law of multiple proportions. According to Lakatos (1971)) for an inductivist, "... only those propositions can be accepted into the body of science which either describe hard facts or are infallible inductive generalizations from them" (p. 92).

In the present case, the experimentally determined "gravimetric combining proportions" would constitute the hard facts (for details see Niaz 2001b).

(b) *Lakatosian* (L): This law was not induced from experimental results but was derived from Dalton's atomic theory and then tested by experiments.

(c) *No mention* (N): Textbook makes no mention explicitly to either of the two interpretations mentioned above.

Results obtained (see Table 4.1) show that 4 textbooks were classified as Inductivist (I), 13 as Lakatosian (L), and 15 as No mention (N). At this stage, it is important to note that this criterion was also used in a previous study (Niaz 2001b) to evaluate 27 general chemistry textbooks published between 1969 and 1999, in the USA. In the present study, 32 general chemistry textbooks published between 1999 and 2014 in the USA were evaluated. A comparison of the two studies could provide some information with respect to any change in the philosophical perspectives of textbooks. The following are examples of textbooks that were classified as Inductivist (I):

> The law of constant composition and the law of multiple proportions were some of the observations that led to the atomic theory of the elements. (McQuarrie et al. 2011, p. 51)

> If two elements form more than one compound, the masses of one element that combine with a fixed mass of the other element are ratios of small whole numbers. (This generalization was first proposed by Dalton early in the nineteenth century and is called the law of multiple proportions.) (Umland and Bellama 1999, p. A35)

Now let us compare these inductivist presentations with those of textbooks that were classified as Lakatosian. The idea of naming these as Lakatosian was simply to establish a clear difference between the inductivist presentations and those that emphasized the role of theories in making predictions. The following are some of the examples of textbooks that were classified as Lakatosian (L):

> Strong support for Dalton's theory came when Dalton and other scientists studied elements that are able to combine to give two (or more) compounds … The law of multiple proportions was not known before Dalton presented his theory. It was discovered because the theory suggested its existence … it was one of the strongest arguments in favor of the existence of atoms. (Brady et al. 2000, p. 49)

> A good theory should not only explain known facts and laws but also predict new ones. The law of multiple proportions was deduced by Dalton from his atomic theory … The deduction of the law of multiple proportions from atomic theory was important in convincing chemists of the validity of the theory. (Ebbing and Gammon 2012, pp. 44–45)

> The hallmark of a good theory is that it suggests new experiments, and this was true of the atomic theory. Dalton realized that it predicted a law that had not yet been discovered. If compounds are formed by combining atoms of different elements on the nanoscale, then in some cases there might be more than a single combination. An example, is carbon monoxide and carbon dioxide … Dalton called this <u>law of multiple proportions</u>, and he carried out quantitative experiments seeking data to confirm or deny it. Dalton and others obtained data consistent with the law of multiple proportions, thereby enhancing acceptance of the atomic theory. (Moore et al. 2011, pp. 22–23, emphasis in the original)

> In 1803, John Dalton proposed a theory to explain the laws of conservation of mass and constant composition. As he developed his atomic theory, Dalton found evidence that required another scientific law that the theory would have to explain … [viz.] law of multiple proportions. (Hill and Petrucci 1999, p. 38, emphasis in the original)

Almost all textbooks classified as Lakatosian showed a considerable understanding of how scientific theories not only explain existing data but also predict new findings. Furthermore, some of these textbooks emphasized (very much in a historical context) that the law of multiple proportions was not known before the theory was postulated and such predictions enhanced the acceptance of Dalton's atomic theory. This result is somewhat surprising, as textbook authors do not generally tend to use historical interpretations. Interestingly, in the present study, 41 % (13 out of 32) of the textbooks were classified as Lakatosian, whereas in the previous study (Niaz 2001b), 26 % (7 out of 27) were classified as such. It is plausible to suggest that given the opportunity and the appropriate circumstances, textbooks do provide interpretation of the development of scientific theories within a historical context.

Criterion D3: Dalton and Gay-Lussac's Law of Combining Volumes

Just as Dalton was working out the details of his atomic theory, Gay-Lussac in 1808 presented his law of combining volumes. For Dalton, to accept Gay-Lussac's law would have amounted to the recognition of the experimental finding that gases combine in simple ratios with respect to their volumes and thus ignore the explanation based on atomic theory. According to Frické (1976), "Gay-Lussac was not alone in rejecting atomism. There was a widespread tendency to replace the theoretical concept of 'atom' by the *measurable notions* of 'volume', 'equivalent', or 'measure'. Dalton's empirical laws, such as the law of multiple proportions, were considered to be of great scientific value, but his *theories were discarded as speculations*" (p. 285, emphasis added; for further details, see Niaz 2001b). The objective of this criterion is to evaluate whether textbooks follow one of the following interpretations with respect to the laws of definite and multiple proportions:

(a) *Inductivist* (I): Gay-Lussac' law of combining volumes provided a rationale for accepting the laws of definite and multiple proportions, without the "superfluous" atomic theory of Dalton. Furthermore, it is suggested that Dalton had not understood Gay-Lussac's and Avogadro's laws, as he did not accept the existence of diatomic molecules.
(b) *Lakatosian* (L): Dalton's atomic theory predicted and partially explained Gay-Lussac's law of combining volumes.
(c) *No mention* (N): Textbook makes no mention explicitly of either of the two interpretations mentioned above.

Results obtained (see Table 4.1) show that 16 textbooks were classified as Inductivist (I), 16 as No mention (N), and none as Lakatosian (L). Following are examples of textbooks that were classified as Inductivist (I):

> Avogadro's principle means that in reactions of gases, their volumes and their number of moles must be in the same ratio. In the equation for the reaction of hydrogen and chlorine, for example, the coefficients are all in the same 1:1:2 ratio as are the numbers of volumes and the number of moles

	$H_2(g) + Cl_2(g) \rightarrow 2\ HCl\ (g)$		
Coefficients	1	1	2
Volumes	1 vol	1 vol	2 vol (experimental)
Molecules (or moles)	1	1	2 (Avogadro's principle)

> *Avogadro's principle* was a remarkable advance in our understanding of gases. His insight enabled chemists for the first time to determine the formulas of gaseous elements … [In a footnote the authors added] … Suppose that hydrogen chloride, for example, is correctly formulated as HCl, not as H_2Cl_2 or H_3Cl_3 or higher … Then the only way that *two* volumes of hydrogen chloride could come from just *one* volume of hydrogen and *one* of chlorine is if each particle of hydrogen and chlorine were to consist of *two* atoms of H and Cl, respectively, H_2 and Cl_2. (Brady et al. 2000, p. 434, italics in the original)

> One hypothesis that was successful in explaining a few of Gay-Lussac's results was that equal volumes of different gases at the same temperature and pressure contained equal number of atoms. John Dalton rejected this idea, however. If two volumes of hydrogen did react with one of oxygen, he reasoned that only *one* volume of steam should have formed, not two … Dalton's reasoning was based on the following *incorrect* equation. $2H\ (g) + O\ (g) \rightarrow H_2O\ (g)$. Avogadro gave the correct explanation of Gay-Lussac's law in 1811. In addition to accepting 'equal volumes-equal numbers' hypothesis, Avogadro proposed that gases may exist in *molecular* form. (Hill and Petrucci 1999, p. 202, italics in the original)

> Unfortunately, other chemists at that time had no appreciation of the difference between atoms and molecules, and could not accept the possibility that elements such as nitrogen and oxygen could consist of diatomic molecules. Avogadro's work was largely ignored for almost 50 years, until Cannizzaro convincingly showed that it leads to a consistent set of atomic weights. (McQuarrie et al. 2011, p. 420)

> John Dalton (who devised the atomic theory) strongly opposed Avogadro's ideas and never did accept them. It took about 50 years—long after Avogadro and Dalton had died—for Avogadro's explanation of Gay-Lussac's experiments to be generally accepted. (Moore et al. 2011, p. 437)

> After publication of the atomic theory .., Dalton assigned a mass of 1 to the hydrogen atom, the lightest known substance. Then, based on the work of Lavoisier, who had shown earlier that water contains 8 g oxygen for every 1 g hydrogen, Dalton assigned a relative mass of 8 to the oxygen atom … At about the same time, the French chemist Joseph Gay-Lussac (1778–1850) also began to study the atom ratio of water. Rather than measuring, *masses*, however, he measured the *volumes* of hydrogen gas and oxygen gas that react to form water vapor … it was assumed that elements, including hydrogen and oxygen, existed as individual atoms. But, then, how could 2 L of individual H atoms combine with 1 L of individual O atoms to yield more than 1 L of water molecules? One suggestion was that each O atom splits in half as two molecules of water form! If that were the case, it meant that atoms were divisible and that the atomic theory was wrong. Dalton vigorously attacked Gay-

> Lussac's technique and results. In 1811, the Italian physicist Amedeo Avogadro (1776–1856) made two proposals that explained these confusing results ... Even though Avogadro's explanation is correct, it was ignored because it was presented with difficult terminology. Almost 50 years later, his ideas were revived and led to the determination of the correct relative mass of the oxygen atom as 16 ... Dalton's model of the atom had survived its first test. (Silberberg 2000, p. 48, italics in the original)

It is important to note that all 16 textbooks classified as Inductivist (I) devoted considerable space to this topic providing historical details with respect to Dalton, Gay-Lussac, Avogadro, and some even to Cannizzaro (some devoted 2–3 pages with color diagrams of chemical reactions illustrating mass–volume relationships). Silberberg (2000, p. 49) even included a note to explain how many renowned scientists denied the existence of atoms, such as Adolf Kolbe (an eminent organic chemist of the nineteenth century), Ernst Mach, and Wilhelm Ostwald. Some of the interesting features of these presentations from general chemistry textbooks are the following:

(a) The use of the formation of HCl, H_2O, and NH_3 as examples to understand the mass–volume relationships in gaseous reactions.
(b) The presentation by Brady et al. (2000), and of some other textbooks, is based on the supposition that hydrogen chloride has the chemical formula HCl and not H_2Cl_2 or H_3Cl_3. Actually, the determination of the chemical formulae (including HCl) was itself problematic in those days.
(c) Avogadro provided a correct explanation of Gay-Lussac's law of combining volumes for gases.
(d) Dalton rejected Gay-Lussac's law and Avogadro's principle as he did not believe in the existence of diatomic molecules.
(e) Despite Avogadro's contribution, it took almost 50 years for relative atomic masses to be determined and the recognition that gases could exist in molecular form.

Despite some problems with these presentations, the historical context in which Dalton, Gay-Lussac, and Avogadro's contributions are discussed can provide students with a glimpse of the dynamics of scientific progress, which inevitably leads to difficulties and controversies that last for many years.

Next, I will contrast these presentations with the historical evidence, which shows that, first, Gay-Lussac did not share Dalton's research program and, second, that Avogadro's explanation of Gay-Lussac's law of combining volumes was itself problematic. According to Frické (1976)), "It is often said that Avogadro was responsible for the great theoretical advance of considering the elementary gases as being diatomic. This is not true, for the molecules of gases were permitted to have any degree of submolecularity, provided that there was either one atom or an even number of atoms in the molecule" (p. 290). Indeed, according to Avogadro (1811) himself, the actual number of atoms in a molecule was "... exactly what is necessary to satisfy the volume of the resulting gas" (reproduced in Frické 1976, p. 290). For example, if six volumes of steam had been produced, Avogadro would have described the reaction as $2\ H_6 + O_6 \rightarrow 6\ H_2O$.

At this stage, it would be of help to widen our historical perspective by including the contributions of the Swedish chemist Jöns Jacob Berzelius (1779–1848), who tried to reconcile the gravimetric (Dalton) and volumetric (Gay-Lussac) data: "What in one theory is called an *atom* is in the other theory a *volume*. In the present state of our knowledge the theory of volumes has the advantage of being founded upon a well-constituted fact, while the other has only a supposition for its foundation" (reproduced in Brock 1993, italics in the original). Given the positivist milieu of the nineteenth century, the interpretation of Berzelius was considered to be more acceptable in the scientific community. Nevertheless, Brock (1993), an eminent historian, has clarified the reconciliation in cogent terms:

> At this point most histories of chemistry point out that, despite Berzelius's impressive exploitation of his knowledge of chemical reactions, the answer to the determination of molecular formulae from which atomic weights could be easily calculated was a hypothesis proposed by Amedeo Avogadro (1776–1856) in 1811, that equal volumes of gases contained the same number of molecules, the latter being stable, multi-atomed particles. In point of fact Avogadro's hypothesis was without any impact or influence on the calculation of atomic weights at this time. Not until the explanatory power of electrochemical theory had temporarily waned in the 1850s under the weight of difficulties in organic chemistry, and chemists and physicists found it convenient to accept (without explanation) that dimers such as H_2 and O_2 could exist, was a complete reconciliation of gravimetric and volumetric data possible. Until then the dimerization of like-charged atoms remained impossible. (p. 165)

Interestingly, the electrochemical theory that Brock refers to is precisely the one postulated and defended by Berzelius in 1804. The electrochemical theory explained the production of electric current in electrolysis as a consequence of chemical decomposition. Furthermore, the historical accounts of Brock (1993) and Frické (1976) agree with each other to a considerable extent and differ from those of general chemistry textbooks. Interestingly, in the present study, 16 (50 %) textbooks were classified as Inductivist (I), whereas in the previous study (Niaz 2001b) only two (7 %) textbooks were classified as such. Table 4.1 also shows that five textbooks had a score of five points (out of a maximum of six points) and the mean score (points) of the textbooks on criteria D1, D2, and D3 was 2.8. Although one may not necessarily agree with the Inductivist (I) interpretation, it did allow for the inclusion of considerable amount of historical content in textbooks.

Presentation of Atomic Models of Thomson, Rutherford, and Bohr in General Chemistry Textbooks

The history of the structure of the atom since the late nineteenth and early twentieth century shows that the atomic models of J. J. Thomson, E. Rutherford, N. Bohr, and Bohr–Sommerfeld evolved in quick succession and had to compete with models based on rival research programs. The emergence of these models shows an underlying pattern that can help to understand the nature of science, in particular the tentative nature of scientific knowledge (for details about the nature of science and its importance for science education, see Chap. 3).

It is plausible to suggest that the evaluation of textbooks based on criteria derived from a history and philosophy of science framework can provide students and teachers with insights as to how models or theories develop and why they change. Ignoring such historical reconstructions can deprive students from an opportunity to familiarize themselves with scientific progress and practice. Furthermore, according to Schwab (1962, 1974), it is important to understand not only the experimental details but also the heuristic principles that underlie the experimental findings. Monk and Osborne (1997) pointed out how many science curricula have forgotten Schwab's important epistemological distinction between the methodological (experimental) and interpretative (heuristic principles) components. Similarly, Matthews (1994) has emphasized the importance of heuristic principles in scientific inquiry and science education in similar terms (for details about heuristic principles, see Chap. 3).

Based on a historical reconstruction of the development of models of atomic structure, Niaz (1998b)) has presented the following criteria for evaluating science textbooks (T = Thomson, R = Rutherford, B = Bohr):

Criterion T1: Thomson's experiments to understand cathode rays as charged particles or waves in the ether

J. J. Thomson's experiments were conducted against the backdrop of a conflicting framework, and he explicitly pointed out that his experiments were conducted to clarify the controversy on the nature of the cathode rays, that is, whether it was charged particles or waves in the ether. This criterion is based on material drawn from Achinstein (1991), Falconer (1987), and Thomson (1897).

Criterion T2: Thomson determined mass-to-charge ratio to decide whether cathode rays were ions or a universal charged particle

Thomson decided to measure mass-to-charge ratio to identify cathode rays as ions (if the ratio was not constant) or as a universal charged particle (constant ratio for all gases). This criterion is based on Achinstein (1991), Heilbron (1964), Niaz (1994), and Thomson (1897).

Criterion R1: Rutherford's nuclear atom

Rutherford's experiments with alpha particles and the resulting model of the nuclear atom had to compete with a rival framework, namely, Thomson's model of the atom (referred to as "plum pudding" in most textbooks). This criterion is based on Niaz (1994) and Rutherford (1911).

Criterion R2: Rutherford argued that the probability of large deflections is exceedingly small, as the atom is the seat of an intense electric field

The crucial reason that clinched the argument in favor of Rutherford's model was not the large-angle deflection of alpha particles (an important finding) but rather the knowledge that one in 20,000 particles deflected through large angles. This criterion is based on Herron (1977), Millikan (1947), and Rutherford (1911).

Criterion R3: Interpretation of alpha-particle experiments as single/compound scattering of alpha particles

To maintain his model of the atom and to explain large-angle deflections of alpha particles, Thomson put forward the hypothesis of compound scattering (multitudes of small scatterings). The rivalry between Rutherford's hypothesis of single scattering based on a single encounter and Thomson's hypothesis of compound scattering led to a bitter dispute between the proponents of the two hypotheses. This criterion is based on Crowther (1910), Rutherford (1911), and Wilson (1983).

Criterion B1: Paradoxical stability of the Rutherford model of the atom

Bohr's main objective was to explain the paradoxical stability of the Rutherford model of the atom, which constituted a rival framework for his own model. This criterion is based on Bohr (1913), Heilbron and Kuhn (1969), Lakatos (1970), and Niaz (1994).

Criterion B2: Explanation of the hydrogen line spectrum

Bohr had not even heard of the Balmer and Paschen formulae for the hydrogen line spectrum, when he wrote the first version of his 1913 article. Failure to understand this episode within a historical perspective led to an inductivist/positivist interpretation, referred to as the "Baconian inductive ascent" by Lakatos (1970). Interestingly, Kuhn and Lakatos, in spite of their so many differences, agreed that Bohr's major contribution was the quantization of the Rutherford model of the atom. This criterion is based on Bohr (1913), Heilbron and Kuhn (1969), Lakatos (1970), and Niaz (1994).

Criterion B3: Deep philosophical chasm

Bohr's incorporation of Planck's "quantum of action" to the classical electrodynamics of Maxwell represented a strange "mixture" for many of Bohr's contemporaries and philosophers of science. This episode illustrates how scientists, when faced with difficulties, often resort to such contradictory "grafts." This criterion is based on Bohr (1913), Holton (1986), Lakatos (1970), and Margenau (1950).

Based on these criteria, textbooks published in the following countries were evaluated:

(a) General chemistry (university-level) textbooks published in the USA ($n=23$, published between 1969 and 1992) and reported in Niaz (1998b).
(b) General chemistry (high school) textbooks published in Venezuela ($n=27$, published between 1972 and 2002) and reported in Páez et al. (2004). These textbooks were published in Spanish.
(c) General chemistry (university-level) textbooks published in Turkey ($n=21$, published between 1964 and 2006) and reported in Niaz and Coştu (2009). These textbooks were published in Turkish.

(d) General physics (university-level) textbooks published in Korea ($n=16$, published between 1992 and 2009) and reported in Niaz et al. (2013). These textbooks were published in Korean.

The following classifications were generated to evaluate the textbooks:
Satisfactory (S): Treatment of the criterion in the textbook is considered to be satisfactory if the role of conflicting frameworks based on competing models of the atom is briefly described.
Mention (M): A simple mention of the conflicting frameworks or controversy is presented with no details.
No mention (N): No mention of the conflicting framework is presented.

Evaluations of textbooks in all four studies were based on inter-rater agreements (for details, see the individual studies). A comparison of the results obtained (see Table 4.2) shows that on all eight criteria a majority of the textbooks were classified as No mention (N). Of the textbooks published in the USA (Niaz. 1998b), none had a Satisfactory (S) presentation on three criteria: T1, R3, and B2. Textbooks published in Venezuela (Páez et al. 2004) were high school textbooks and apparently had more difficulty in including material related to history and philosophy of science as none of the textbooks had a Satisfactory (S) presentation on seven criteria: T2, R1, R2, R3, B1, B2, and B3.

In the case of textbooks published in Turkey (Niaz and Coştu 2009), none had any Satisfactory (S) presentation on four criteria: T1, R3, B2, and B3. In the case of Korea (Niaz et al. 2013), general physics textbooks were analyzed (as we could not contact colleagues who teach chemistry), and none had a Satisfactory (S) presentation on four criteria: T1, T2, R3, and B2. Based on these results, it is plausible to suggest that textbooks published in four different cultures, in different continents,

Table 4.2 Comparison of the presentation (percentages) of atomic structure in textbooks published in the USA, Venezuela, Turkey, and Korea based on history and philosophy of science criteria

Criteria	USA ($n=23$)			Venezuela ($n=27$)			Turkey ($n=21$)			Korea ($n=16$)		
	N	M	S	N	M	S	N	M	S	N	M	S
T1	91	9	–	93	4	4	100	–	–	100	–	–
T2	91	–	9	93	7	–	71	24	5	94	6	–
R1	52	17	30	100	–	–	19	10	71	50	25	25
R2	91	–	9	81	19	–	38	52	10	50	38	13
R3	100	–	–	100	–	–	100	–	–	100	–	–
B1	70	17	13	78	22	–	62	10	29	25	–	75
B2	100	–	–	100	–	–	100	–	–	13	88	–
B3	74	17	9	100	–	–	100	–	–	88	6	6

Notes:
1. Results in this table are reproduced from the USA (Niaz 1998b), Venezuela (Páez et al. 2004), Turkey (Niaz and Coştu 2009), and Korea (Niaz et al. 2013)
2. N=No mention, M=Mention, S=Satisfactory (for details see text)
3. Figures under N, M, and S represent percentages
4. Criteria T1, T2, R1, R2, R3, B1, B2, and B3: see text for details

have an underlying thread, namely, the dominant empiricist epistemology, in which atomic models are based almost entirely on experimental data.

Although most of the textbooks ignore the changing nature of scientific theories (Niaz 2009), some textbooks did provide thought-provoking presentations that were classified as Satisfactory (S). The following is an example of a textbook from Turkey that was classified as Satisfactory (S) on criterion R2:

> As seen in Figure… it was observed that most of the particles pass through with no deflection, however, one alpha particle in 20,000 was deflected through an angle greater than 90°. Rutherford was intrigued by the observations from alpha particle experiments, since if Thomson's model in which the positive charge and the mass are distributed evenly throughout the atom is valid, alpha particles must pass through with little deflection and not scattered backwards…[from his observation] Rutherford concluded that:
>
> 1. Since most of the alpha particles pass through with no deflection, an atom must consist largely of empty space.
> 2. Since few alpha particles, atoms of He with charge +2 [He^{2+}], were deflected through an angle greater than 90°, an atom must consist of a positively charged nucleus the size of which is much smaller as compared to the atom. (Tunalı and Aras 1977, p. 217)

This presentation refers to Rutherford's alpha-particle experiments, and it shows clearly that in order to be classified as Satisfactory (S), it was not necessary to include extra historical details but rather recognize that how, based on experimental results, "Rutherford was intrigued" and how this led him to introduce *changes* in Thomson's model.

Another example is provided from a textbook published in the USA that was classified as Satisfactory on criterion B3:

> There are two ways of proposing a new theory in science, and Bohr's work illustrates the less obvious one. One way is to amass such an amount of data that the new theory becomes obvious and self-evident to any observer. The theory then is almost a summary of the data. The other way is to make a bold new assertion that initially does not seem to follow from the data, and then to demonstrate that the consequences of this assertion, when worked out, explain many observations. With this method, a theorist says, "You may not see why, yet, but please suspend judgment on my hypothesis until I show you what I can do with it." Bohr's theory is of this type. Bohr said to classical physicists: "You have been misled by your physics to expect that the electron would radiate energy and spiral into the nucleus. Let us assume that it does not, and see if we can account for more observations than by assuming that it does." (Dickerson et al. 1984, p. 264)

This example shows how in the face of difficulties, scientists still manage to solve the problems by asking their colleagues to "suspend judgment on my hypothesis" and thus introduce *changes* in the corpus of scientific knowledge. Again, this presentation provides an example as to how textbooks can easily include thought-provoking questions that not only approximate the historical record but also arouse students' curiosity.

Presentation of the Bohr–Sommerfeld Model in General Chemistry Textbooks

In the quest for understanding why atomic models change, we next consider the Bohr–Sommerfeld model of the atom presented in 1915–1916. Bohr's model of the atom provided an explanation of the paradoxical stability of the Rutherford model and spectra of hydrogen-like ions. Despite its popularity and novelty, Bohr's model only explained the stability, ionization energy, and spectra of ions possessing a single electron (H^+, Li^{2+}, Be^{3+}). Sommerfeld's (1915, 1916) contribution consisted in treating the problem relativistically by introducing elliptical orbits, in which the electrons penetrated the region of internal electrons. Thus, the highly elliptical orbits would have additional stability. The Bohr–Sommerfeld model of the atom was widely accepted by the scientific community as an improvement of Bohr's model. For example, Paschen's measurement of the helium spectrum was in agreement with Sommerfeld's prediction. Despite its success and novelty, the Bohr–Sommerfeld model went no further than the spectra of alkali metals, which led scientists to look for other models, and consequently the atomic model had to *change* again. These difficulties were resolved by Pauli's exclusion principle and new developments in quantum mechanics, leading to the postulation of the wave-mechanical model of the atom.

Based on these considerations, it is plausible to suggest that in order to facilitate a better understanding of how scientific models change and the tentative nature of scientific knowledge, it is important for textbooks to include the following aspects:

1. That Bohr's model of the atom could only explain the spectra of hydrogen-like ions, based on circular orbits
2. That the Bohr–Sommerfeld model of the atom based on elliptical orbits, not only specified the shape of the orbit and its orientation in space but also provided additional stability

Based on these aspects, 28 general chemistry textbooks published in Italy (between 1969 and 2008) and 46 published in the USA (between 1968 and 2008) were evaluated by Niaz and Caredellini (2011a, b) and classified in the following categories:

Satisfactory (S): Presentation of a textbook was considered to be "satisfactory" if it included a description of the Bohr–Sommerfeld model along with diagrams of the elliptical orbits.

Mention (M): Presentation of a textbook was considered to be "mention" if it made a simple mention of the model and/or elliptical orbits with no diagrams or details.

No mention (N): Textbooks in this category made "no mention" to the Bohr–Sommerfeld model or elliptical orbits.

Textbooks published in Italy were included in order to have a different cultural perspective of the Bohr–Sommerfeld model of the atom. Furthermore, all Italian textbooks included in this study were not translations of those originally published in the USA. Of the 28 Italian textbooks, 5 were classified as Satisfactory (S), 7 as Mention (M), and 16 as No mention (N). The following results were obtained from the 46 textbooks published in the USA: Satisfactory (S) = 3, Mention (M) = 3, and No mention (N) = 40. Despite the similarities, the Italian textbooks presented the Bohr–Sommerfeld model better than those published in the USA. The following is an example of an Italian textbook that was classified as Satisfactory (S):

> Sommerfeld, in 1916 made a first important improvement of Bohr's model, by introducing a second quantum number based on the previous general considerations [the analogy with the planets' motion]. If the orbits are elliptical, the position of the electron in the orbital plane is defined by two periodic quantizable variables, the radius vector r and the angle φ (fig. 3). A double quantization carries then along to define two quantum numbers .... (Lorenzelli 1969, p. 27, in a section entitled "The secondary quantum number l and the Sommerfeld's atom")

The following is an example of a textbook published in the USA that was classified as Satisfactory (S):

> Arnold Sommerfeld (1868–1951) proposed an ingenious way of saving the Bohr theory. He suggested that orbits might be elliptical as well as circular. Furthermore, he explained the differences in stability of levels with the same principal quantum number, n, in terms of the ability of the highly elliptical orbits to bring the electron closer to the nucleus (Figure 7–15). For a point nucleus of charge +1 in hydrogen, the energies of all levels with the same n would be identical. But for a nucleus of +3 screened by an inner shell of two electrons in Li, an electron in an outer circular orbit would experience a net attraction of +1, whereas one in a highly elliptical orbit would penetrate the screening shell and feel a charge approaching +3 for part of its traverse. Thus, the highly elliptical orbits would have the additional stability … The *s* orbit, being the most elliptical of all in Sommerfeld's model, would be much more stable than the others in the set of common n … The Sommerfeld scheme led no further than the alkali metals. Again an impasse was reached, and an entirely fresh approach was needed. (Dickerson et al. 1984, pp. 269–271, italics in the original, in a section entitled "Need for a better theory")

These presentations clearly show the need for *changes* (improvements) in Bohr's model of the atom and go further by suggesting elliptical orbits as an alternative in order to provide greater stability and thus provide a better explanation of atomic spectra. Dickerson et al. (1984) explicitly discuss how in the elliptical orbits an electron can penetrate the "screening shell," approach the nucleus, and thus achieve greater stability. Furthermore, these authors point out the shortcomings of the Bohr–Sommerfeld model and how that led to another "impasse" and hence the need for further changes. This was eventually accomplished by the postulation of the wave-mechanical model of the atom. This study also showed that most general chemistry textbooks ignore the Bohr–Sommerfeld model and very few consider it as a manifestation of the tentative nature of scientific theories. It is plausible to suggest that the following sequence of atomic models, Thomson → Rutherford → Bohr → Bohr–Sommerfeld and subsequently wave mechanical, constitutes an illustration of how models change and why no model can be considered to be perfect and thus last forever. In the case of these atomic models, this is even more understandable as these *changes* occurred over a relatively short period of about 20 years.

Revisiting the Presentation of Atomic Models of Thomson, Rutherford, and Bohr

Given the importance and widespread coverage of these models in textbooks published in various parts of the world, I decided to evaluate their presentation in more recently published general chemistry textbooks. In the previous study (Niaz 1998b) reported in an earlier section of this Chap., 23 general chemistry textbooks published in the USA, between 1969 and 1992, were evaluated. In the present study, published for the first time, 39 general chemistry textbooks published in the USA, between 1995 and 2014, were evaluated. It would be of interest to see if the textbooks have improved their presentations (as compared to the previous study by Niaz 1998b) based on history and philosophy of science-related criteria used in the previous study. Results obtained on the same eight criteria (T1, T2, R1, R2, R3, B1, B2, and B3) are presented in Table 4.3. At this stage, it is of interest to compare the mean score (points) of the 39 textbooks in the present study (1.59) with that of the previous study (2.0) by Niaz (1998b)). This decrease is somewhat disappointing. However, a few textbooks in this study presented some historically based presentations of very high quality.

Table 4.3 Evaluation of atomic structure in general chemistry textbooks based on a history and philosophy of science framework

	Criteria[a]								
Textbook	T1	T2	R1	R2	R3	B1	B2	B3	Points[b]
1. Atkins and Jones (2002)	N	N	M	S	N	N	N	N	3
2. Atkins and Jones (2008)	N	N	M	S	N	N	N	N	3
3. Bettelheim et al. (2012)	N	N	N	N	N	N	N	N	0
4. Bishop (2002)	N	N	N	N	N	N	N	N	0
5. Brady and Humiston (1996)	N	N	N	N	N	N	N	N	0
6. Brady et al. (2000)	N	N	N	N	N	N	N	N	0
7. Brown et al. (1997)	N	N	M	N	N	N	N	N	1
8. Brown et al. (2012)	M	N	M	N	N	N	N	N	2
9. Burns (1996)	N	N	N	N	N	N	N	N	0
10. Chang (1998)	N	N	N	N	N	N	N	N	0
11. Chang (2010)	N	N	N	N	N	M	N	N	1
12. Daub and Seese (1996)	N	N	N	N	N	N	N	N	0
13. Denniston et al. (2011)	N	N	N	N	N	N	N	N	0
14. Dickson (2000)	N	N	N	N	N	N	N	N	0
15. Ebbing (1996)	N	N	N	S	N	M	N	N	3
16. Ebbing and Gammon (2012)	N	N	N	S	N	M	N	N	3
17. Frost et al. (2011)	N	N	N	N	N	N	N	N	0
18. Goldberg (2001)	N	N	N	N	N	N	N	N	0
19. Hein and Arena (1997)	N	N	N	N	N	N	N	N	0
20. Hill and Petrucci (1999)	S	S	M	S	N	N	N	N	7

(continued)

Table 4.3 (continued)

	Criteria[a]								
Textbook	T1	T2	R1	R2	R3	B1	B2	B3	Points[b]
21. Jones and Atkins (2000)	N	N	M	S	N	N	N	N	3
22. Malone (2001)	N	N	M	N	N	M	N	N	2
23. McMurry et al. (2007)	N	N	N	N	N	N	N	N	0
24. McMurry and Fay (2001)	N	N	N	S	N	N	N	N	2
25. McQuarrie et al. (2011)	N	N	N	N	N	N	N	N	0
26. Moore et al. (2002)	N	N	N	N	N	N	N	N	0
27. Moore et al. (2011)	N	N	N	N	N	N	N	N	0
28. Oxtoby et al. (2012)	M	N	S	S	N	M	N	N	6
29. Raymond (2010)	N	N	N	N	N	N	N	N	0
30. Russo and Silver (2002)	N	N	M	S	N	N	N	N	3
31. Seager et al. (2011)	N	N	N	N	N	N	N	N	0
32. Silberberg (2000)	N	S	M	S	N	N	N	M	6
33. Spencer et al. (2012)	N	N	N	S	N	N	N	N	2
34. Stoker (2010)	N	N	N	N	N	N	N	N	0
35. Timberlake (2010)	N	N	M	N	N	N	N	N	1
36. Tro (2008)	N	N	N	S	N	M	N	N	3
37. Umland and Bellama (1999)	N	N	N	S	N	M	N	N	3
38. Whitten et al. (2013)	N	M	M	S	N	N	N	N	4
39. Zumdahl et al. (2014)	N	N	S	N	N	S	N	N	4

Notes:
References to textbooks in this table are presented in Appendix 1
[a]Criteria: T = Thomson, R = Rutherford, B = Bohr, S = Satisfactory, M = Mention, N = No mention
T1: Cathode rays as charged particles or waves in the ether
T2: Cathode rays as ions or universal charged particles
R1: Rutherford's nuclear atom versus Thomson's model of the atom
R2: Large-angle deflection versus deflection of one in 20,000 particles
R3: Single/compound scattering of alpha particles
B1: Paradoxical stability of the Rutherford model of the atom
B2: Bohr's explanation of the hydrogen line spectrum
B3: Incorporation of Planck's "quantum of action"– a deep philosophical chasm
[b]Points: S = 2 points, M = 1 point, N = 0 point

Thomson's Model (Criteria T1 and T2)

The following is an example of a textbook that was classified as Satisfactory (S) on criterion T1, which referred to the nature of the cathode rays (viz., charged particles or waves in the ether):

> Although German and British scientists agreed on the facts, they developed strikingly different hypotheses about cathode rays. Most of the German scientists thought that the rays were a form of electromagnetic radiation, much like light. Most British scientists thought that the rays were particles of matter—probably residual gas molecules that had acquired a negative charge from the cathode. When J.J. Thomson assessed all the data on cathode rays available in 1897, he leaned heavily toward the particle hypothesis. Thomson then settled

the question unequivocally in a landmark series of experiments. (Hill and Petrucci 1999, p. 277)

This clearly shows that there is a reason and a rationale based on which a scientist designs his/her experiments. Let us compare this presentation with that of a textbook that was classified as No mention (N):

> In the mid-1800s, scientists began to study electrical discharge through partially evacuated tubes … These observations of the properties of cathode rays suggested that the radiation consists of a stream of negatively charged particles, which we now call *electrons*. In addition, it was found that the cathode rays emitted by different cathode materials were the same. All of these observations led to the conclusion that electrons are a basic component of matter. (Brown et al. 1997, p. 39)

This presentation clearly lacks the understanding that the cathode-ray experiments were conducted against the backdrop of a conflicting framework, namely, electromagnetic radiation or particles of matter. Furthermore, it gives the impression that the experiments directly led to the postulation of the electrons and thus ignores the background to the controversy that helped scientists to understand the experimental results. It is generally believed that the electron was discovered in 1897 by J. J. Thomson. Arabatzis (2006) questions this claim and based on his biographical approach to representing electrons argues convincingly that no historical episode where controversy persists can be interpreted as constituting a discovery. Besides Thomson's work on cathode rays, there were many other experimental situations that were interpreted as observable manifestations of the electron, such as the Zeeman effect, the scattering of alpha and beta particles, the photoelectric effect, the cloud chamber tracks, and Millikan's oil drop experiments. Consequently, the electron was not discovered in 1897 by Thomson, but rather its representation was enriched through the collective efforts of various scientists over many years until a consensus was achieved in the scientific community.

The following is an example of a textbook that was classified as No mention (N) on criterion T2 that refers to the determination by Thomson of the mass-to-charge ratio of cathode rays:

> By adjusting the charge on the electrodes, … Thomson was able to calculate the first bit of quantitative information about a cathode ray particle—the ratio of its charge to its mass … Many experiments were performed using the cathode ray tube, and they demonstrated that cathode ray particles are in all matter. They are, in fact, *electrons*. (Brady et al. 2000, p. 57)

Many textbooks provided considerable details with respect to the experimental determination of the mass-to-charge ratio of cathode rays (e.g., McMurry and Fay 2001; Oxtoby et al. 2012). However, these textbooks do not explain how this helped Thomson to reach the conclusion that cathode rays were universal charged particles. This is the crux of the issue: Thomson reasoned that if the ratio was constant for different gases, it would provide evidence for a universal charged particle. In contrast, if the ratio was different for each gas, that would indicate that cathode rays were charged ions.

Textbooks that describe all the experimental details with no reference to Thomson's reasoning for determining mass-to-charge ratio will probably leave the students perplexed and even perhaps disconcerted as to how the data helped him to reach his conclusion. Holton (1969) has described such presentations as the "experimenticist fallacy." In other words, according to Holton experiments are performed with a particular purpose that helps the scientists to obtain evidence for or against a particular hypothesis. Similarly, such presentations in textbooks have been referred to by Schwab (1962, 1974) as a "rhetoric of conclusions," namely, enumeration of experimental details with no reference to the reasoning or rationale of the experiment.

Now, let us compare such presentations (experimenticist/rhetoric) with that of a textbook that was classified as Satisfactory (S):

> In another set of experiments, he [Thomson] showed that the magnetic deflections were the same no matter what the residual gas in the cathode ray tube—whether it was hydrogen, air, carbon dioxide, or other gases. This observation strongly suggested that cathode rays are not ions formed from gaseous atoms or molecules; instead, they are negatively charged particles *found in all matter*. To strengthen his argument, Thomson designed an experiment to obtain an easily measured property of cathode rays: the ratio of their mass (m_e) to charge e. (Hill and Petrucci 1999, pp. 277–278, italics in the original)

In order to pursue this aspect of Thomson's experiments further, Niaz et al. (2002) conducted a study within a history and philosophy of science framework in which general chemistry students were asked the following question: How would you have interpreted, if on using different gases in the cathode-ray tube (Thomson's experiment), the relation *e/m* [charge to mass] would have resulted different? These students formed part of an experimental group in the study in which they were encouraged to argue, discuss, and resolve contradictions. If experiments are performed to have evidence for or against a particular hypothesis, then this was a difficult question, as the students had already learned in their course that Thomson had found the charge-to-mass ratio (*e/m*) to be constant. Despite this difficulty, 37 % of the students understood correctly the implications of this hypothetical experimental finding, and the following are two examples: "a) Different gases would not have a universal particle, and the relation *e/m* would depend on the gas used in the experiment; b) Each gas would contain different particles … in other words, the fundamental particle with a constant value of *e/m* independent of the gas used, would not exist" (reproduced in Niaz et al. 2002, p. 521). Interestingly, 49 % of the students gave a rhetorical response with no justifications, such as that Thomson had already found that the ratio *e/m* was constant. This clearly shows the need for exploring the complex relationship between experiment and theory within a domain-specific context of the science curriculum.

Rutherford's Model (Criteria R1, R2, and R3)

Criterion R1 dealt with the postulation of the nuclear atom based on Rutherford's alpha-particle experiments. Here is an example of a textbook that was classified as Satisfactory (S):

> Consider Ernest Rutherford's alpha-particle bombardment experiment illustrated in Fig. 2.12. How did the results of this experiment lead Rutherford away from the plum pudding model of the atom to propose the nuclear model of the atom? (Zumdahl and Zumdahl 2014, p. 72)

This textbook correctly invites students (even arouses their curiosity) to think and reflect with respect to new experimental evidence and how it can change an existing model (Thomson) and subsequently postulate a new one (Rutherford). Now, contrast this example with the presentations of the following textbooks:

> [Thomson's] model was overthrown in 1908 by another experimental observation. (Atkins and Jones 2008, p. 3)

> In 1908, this 'plum-pudding' model was overthrown by a simple experiment. (Jones and Atkins 2000, p. 7)

> What ultimately banished this model to trash heap was an experiment performed just a few years later, in 1909, by the British physicist Ernest Rutherford. (Russo and Silver 2002, p. 81)

An understanding of atomic models based on a history and philosophy of science perspective would not consider a model that is replaced by another as "overthrown" or "banished to trash heap." On the contrary, it is preferable to acknowledge the role (heuristic power) of an existing model and then show that the succeeding model has greater explanatory power. In other words, scientific models are not right or wrong, and still *change* is inevitable. For understanding the role of competing/changing models, see Chap. 2.

Criterion R2 dealt with the probability of large-angle deflections in Rutherford's alpha-particle experiments. Here are examples of textbooks that were classified as Satisfactory (S):

> What would be a feasible model for the atom if Geiger and Marsden had found that 7999 out of 8000 alpha particles were deflected back at the alpha-particle source. (Ebbing and Gammon 2012, p. 48)

> (a) The observations: (1) Most of the alpha particles pass through the foil undeflected. (2) Some alpha particles are deflected slightly as they penetrate the foil. (3) A few (about 1 in 20,000) are largely deflected. (4) A similar small number do not penetrate the foil at all, but are reflected back toward the source.
>
> (b) Rutherford's interpretation: If the atoms of the foil have a massive, positively charged nucleus and light electrons outside the nucleus, one can explain how: (1) an alpha particle passes through the atom undeflected (a fate shared by most of the particles); (2) an alpha particle is deflected slightly as it passes near an electron; (3) an alpha particle is strongly deflected by passing close to the atomic nucleus; and (4) an alpha particle bounces back as it approaches the nucleus head-on. (Hill and Petrucci 1999, p. 282)

> These deflections were surprising, but the 0.001 % of the total that were deflected at acute angles … were totally unexpected. (Whitten et al. 2013, p. 120)

In the presentation of the alpha-particle experiments, Ebbing and Gammon (2012) stated that one in 8000 alpha particles deflected back through large angles. This led Rutherford to postulate the existence of the nucleus with an intense electric field. Consequently, it is interesting and thought provoking to ask students to consider a different experimental result in which 7999 of the 8000 alpha particles were deflected. This would necessarily require a different atomic model from the one proposed by Rutherford. The relationship between observations and models or theories is not as straightforward as many textbook authors seem to imply. History of science shows that experimental data generally do not dictate the theory. Elaboration of a model or theory is a complex task. In this respect, the presentation by Hill and Petrucci (1999) is interesting and goes beyond that of Ebbing and Gammon (2012) by clearly differentiating between observations and Rutherford's interpretations. This is an attempt to explain the different types of deflections based on the hypothesis that there exists a "massive positively charged nucleus." Interestingly, Thomson also did the alpha-particle experiments and found very similar experimental results. Yet, his interpretations were entirely different and hence the ensuing controversy (this is the subject of criterion R3).

Another textbook adopted a novel approach by posing the following questions in a section entitled "Critical thinking":

> You have learned about three different models of the atom: Dalton's model, Thomson's model, and Rutherford's model. What if Dalton was correct? What would Rutherford have expected from his experiments with gold foil? What if Thomson was correct? What would Rutherford have expected from his experiments with gold foil. (Zumdahl and Zumdahl 2014, p. 53; Similarly, the authors also show the transition from Bohr's model to the quantum mechanical model)

These are interesting questions and make a lot of sense in classroom discussions. If Dalton's model were correct, then Rutherford's expectations from alpha-particle experiments would have been different. Similarly, if Thomson was correct, then Rutherford's interpretations could be challenged. Actually, this is what happened as Thomson thought that his model could also explain the large-angle deflections of alpha particle. (The controversy between Thomson and Rutherford lasted for many years. See criterion R3 and for further details see Niaz 1998b and Wilson 1983.)

In order to pursue this aspect of Rutherford's experiments and the ensuing Thomson–Rutherford controversy further, Niaz et al. (2002) conducted a study within a history and philosophy of science framework in which general chemistry students were asked the following question: If Rutherford's experiments changed Thomson's model of the atom entirely, in your opinion did Thomson make mistakes while doing his experiments? These students formed part of an experimental group in the study in which they were encouraged to argue, discuss, and resolve contradictions. Interestingly, 62 % of the students responded correctly by explicitly recognizing that Thomson did not make mistakes and besides that Rutherford's model improved upon the previous findings. Here are three examples of how these students reasoned: a) "In science, one experiment leads to another with the objective of improving on previous findings … In the case of Thomson and Rutherford, Rutherford's model provided further information with respect to the atomic structure"; b) "Thomson in a way elaborated on the ideas of Democritus by experimental means … but as everything in science is subject to modification, Rutherford elaborated on Thomson's model"; c) "The fact that Thomson's model had many deficiencies does not mean that we should not recognize his merits … Thomson's work motivated others (e.g., Rutherford) to keep investigating about the structure of the atom" (Reproduced in Niaz et al. 2002, pp. 521–522). Given the present state of textbooks, the following aspects of these responses are highly encouraging and motivating for future studies: in science one experiment leads to another. Thomson elaborated on the ideas of Democritus experimentally. Although the relationship between Democritus and Thomson may seem far-fetched, it is at present the subject of debate among historians and philosophers of science. However, a word of caution is necessary as in this study 14 % of the students responded that Thomson had made mistakes while doing his experiments.

Criterion R3 dealt with two rival hypotheses to explain the findings of the alpha-particle experiments, namely, Rutherford's *hypothesis* of *single scattering* and Thomson's *hypothesis of compound scattering*. Interestingly, none of the textbooks in this study (or for that matter any general chemistry textbooks in the previous studies) had a Satisfactory (S) presentation on this criterion. However, two general physics textbooks published in the USA in a previous study (Rodríguez and Niaz 2004) were classified as Satisfactory (S), and the relevant parts of the text are reproduced here:

> The Thomson model for scattering fails when we examine the probability for scattering at large angles. If each individual scattering deflects the projectile through an angle of around 0.01°, then to observe projectiles scattered through a total angle greater than 90°, we must have about 10^4 successive scatterings, *all* of which push the projectile toward larger angles. Since the probabilities of individual scatterings toward either larger or smaller angles are equal, the probability of having 10^4 successive scatterings toward larger angles, like the probability of finding 10^4 successive heads in tossing a coin, is about $(1/2)^{10.000} = 10^{-3000}$. (Krane 1996, p. 178)

> Rutherford calculated that from the large Thomson positive charge distribution particles should never be deflected more than 0.03 degrees in a single collision; in undergoing multiple collisions they should have about an equal chance of being deflected one way as another. Therefore, large deflections as a result of many single deflections in the same direction were very improbable. (It had been calculated on the basis of the Thomson model that a total deflection greater than 90° in traversing the gold foil would have only one chance in 10^{3500} of occurring.) (Cooper 1970, p. 321)

Indeed, both presentations argue persuasively against Thomson's hypothesis of compound scattering and at the same time provide arguments to support Rutherford's hypothesis of single scattering. To facilitate understanding, Krane provides an analogy with the tossing of a coin and thc probability of finding successive heads. Cooper argues that there was very low probability of many single deflections in the same direction. At this stage, it would be interesting to ask, as to how did Krane and Cooper come to have arguments based on the probability of the deflections. As is generally the case, history of science is a rich source of information provided that we look at the historical evidence. Rutherford (1911) himself in his seminal article provided the arguments based on probability in the following terms:

> A simple calculation based on the theory of probability shows that the chance of an alpha particle being deflected through 90° is vanishingly small. In addition, it will be seen later that the distribution of the alpha particles for various angles of large deflexion does not follow the probability law to be expected if such large deflexions are made up of a large num-

> ber of small deviations. It seems reasonable to suppose that the deflexion through a large angle is due to a single encounter, for the chance of a second encounter of a kind to produce a large deflexion must in most cases be exceedingly small. (p. 669)

In a sense, both Krane and Cooper are reiterating what Rutherford himself presented as an argument to discard the deflections of alpha particles for large angles based on probability distributions, especially if these are made up of a large number of small deviations as suggested by Thomson. Interestingly, early in 1909, Rutherford had enrolled to attend elementary lectures on probability given by Horace Lamb, and his notebooks bear witness to his attendance and to having taken extensive notes (Wilson 1983, p. 290). This leads us to yet another question: As both Krane and Cooper are not historians or philosophers of science, so how did they manage to include arguments related to probability theory, as employed by Rutherford? I could not trace the origin of Krane's ideas. Cooper (1968), however, does mention that he has profited from some historical writings:

> I have profited particularly from Lane Cooper's *Aristotle, Galileo, and the Tower of Pisa*, Charles Coulston Gillispie's *The edge of objectivity*, William Francis Magie's *A source book in physics*, Henry A. Boorse and Lloyd Motz's *The world of the atom*, the Physical Science Study Committee text, *Physics*, Aaron Bork's *Foundations of electromagnetic theory: Maxwell*, and Thomas S. Kuhn's *The Copernican revolution*. (Preface, p. vii)

Actually Leon Cooper, a theoretical physicist and an active researcher in cellular and molecular basis for learning and memory, has considerable interest in the history of science. He has authored a number of general physics textbooks in which he presents the different topics within a history and philosophy of science perspective (for details of Cooper's historical perspective on teaching science, see Niaz et al. 2010b.

Bohr's Model (Criteria B1, B2, and B3)

Criterion B1 dealt with the paradoxical stability of the Rutherford model of the atom and with how Bohr solved this problem. Although Bohr's model subsequently also attempted to explain atomic spectra of some elements, his major objective was to improve Rutherford's model (for details, see Niaz 1998b; Wilson 1983). In this study, only one textbook had a Satisfactory (S) presentation and seven textbooks were classified as Mention (M). The following is an example of a textbook that was classified as M:

> The Danish physicist Neils Bohr (1885–1962) attempted to develop a model for the atom that explained atomic spectra. In his model, electrons travel around the nucleus in circular orbits (similar to those of the planets around the sun) … Bohr called these orbits *stationary states* and suggested that, although they obeyed the laws of classical mechanics, they also possessed 'a peculiar, mechanically unexplainable, stability.' … Bohr further proposed that, in contradiction to classical electromagnetic theory, no radiation was emitted by an electron orbiting the nucleus in a stationary state. (Tro 2008, p. 293)

This is a fairly good presentation and could have been classified as Satisfactory (S) if it had emphasized that Bohr's major objective was not the explanation of atomic spectra but rather the "unexplainable stability." Actually, this is found in most general chemistry and physics textbooks of previous studies. It is precisely this "contradiction" in Rutherford's model of the atom that led Bohr to propose the inclusion of the "quantum of action" in order to explain the paradoxical stability, and this constitutes a major turn in the development of atomic structure (also with respect to mechanical instability of Rutherford's atom, see Heilbron and Kuhn 1969). Actually, such "turns" constitute an important part of scientific progress. Interestingly, according to some philosophers of science (e.g., Lakatos 1970), if the rules of inductivism had prevailed, Bohr's (1913) article should not have been published, as it was based on a contradiction. In order to clarify the issue even further, let us compare these presentations with that of a textbook that was classified as No mention (N):

> Although the Bohr theory satisfactorily explained the spectra of hydrogen and of other species containing one electron (He^+, Li^{2+}, etc.) the wavelengths in the observed spectra of more complex species could not be calculated. Bohr's assumption of circular orbits was modified in 1916 by Arnold Sommerfeld (1868–1951), who assumed elliptical orbits. Even so, *the Bohr approach was doomed to failure*, because it modified classical mechanics to solve a problem that could not be solved by classical mechanics. It was a contrived solution. (Whitten et al. 2013, p. 144, italics added)

The reference to Sommerfeld's elliptical orbits in this presentation is interesting and has been the subject of discussion in a previous section of this chapter. Basically, the Bohr–Sommerfeld model was an attempt to increase the empirical success (heuristic power) of Bohr's model. However, the assertion that the *Bohr approach was doomed to failure* is questionable. Actually, history of science shows that in the long run all theories turn out to be "false" and consequently shall we consider all science to be *doomed to failure* (Lakatos 1970, p. 158). The obvious answer to this problem is that this is not the case. This thesis has been endorsed in science education by Burbules and Linn (1991, p. 232), who have argued cogently that history of science shows that in the long run all theories more or less turn out to be "wrong." This precisely shows the progressive nature of science. Otherwise, it would be difficult to understand how Einstein's theory of relativity could have greater explanatory power than Newton's (see Chaps. 2 and 3, for further discussion). Does this mean that Newton's laws were "wrong" or "false"? These are important issues for science education and also show that the increase in the explanatory power of subsequent theories is also a manifestation of the tentative nature of science.

Criterion B2 dealt with the role played by the hydrogen line spectrum (Balmer and Paschen formulae) in the development of Bohr's model of the atom. Actually, Bohr had not even heard of these formulae when he wrote the first version of his 1913 article in the *Philosophical Magazine*. None of the general chemistry textbooks in this study were classified as Satisfactory (S) or Mention (M). The following is an example of a textbook that was classified as No mention (N):

> As seen in Table 7.2, there is exceptional agreement between the experimentally measured wavelengths and those calculated by the Bohr theory. Thus, Niels Bohr had tied the unseen

> (the atom) to the seen (the observable lines of the hydrogen emission spectrum)—a fantastic achievement! (Moore et al. 2011, p. 284)

Such presentations do look impressive and even fantastic! The ahistoric nature of this presentation actually detracts from Bohr's real merit, namely, the quantization of Rutherford's model of the atom in order to solve the problem of the paradoxical stability. A historical reconstruction shows that Bohr was not initially looking for an agreement between the experimental data and his theory. Bohr's theoretical insight opened a whole new world for the study of atomic structure, and the problem of studying the spectra became important later. Actually, such presentations are quite common in textbooks and show how the inductive nature of science is paramount! In order to understand the underlying issues further, let us consider the following presentation from a general physics textbook that was classified as Mention (M) on this criterion:

> If the emission of light by the atom was quantized, which is what the Balmer formula suggested, then the structure of the atom itself had to be quantized in some way too. Bohr's progress was slowed by his unfamiliarity with Balmer's formula. When a friend showed Balmer's work to him in early 1913, everything fell into place; Bohr had the answer. (Cohen 1976, p. 235; reproduced in Rodríguez and Niaz 2004, p. 420)

Indeed, as compared to Moore et al. (2011), this presentation comes closer to the historical record and provides a better appreciation of Bohr's contribution. Nevertheless, it still lacks the understanding that Bohr was not waiting for Balmer's formula in order to postulate his model of the atom and there is no evidence to show that this may have slowed down his progress.

Criterion B3 dealt with Bohr's postulation of Planck's "quantum of action" to the classical electrodynamics of Maxwell as a contradictory "graft." None of the textbooks in this study were classified as Satisfactory (S), and only one was classified as Mention (M), and the following is an excerpt:

> When Niels Bohr … who had become a champion of the new physics, delivered a lecture in Rutherford's laboratory on the principle, Rutherford … said, "You know, Bohr, your conclusions seem to me as uncertain as the premises on which they are built." Acceptance of radical ideas does not come easily, even to fellow geniuses. (Silberberg 2000, p. 276)

This presentation gives students a glimpse of how Bohr's model was contradictory and had difficulties for its acceptance in the scientific community. In order to facilitate further understanding of the difficulties involved with Bohr's model of the atom, let us consider the following presentation from a Korean general physics textbook that was classified as Satisfactory (S):

> After many people realized that the model of [the] solar system is not a good explanation for the structure of [the] atom, they were in total confusion. Meanwhile, Danish physicist, Bohr proposed a new model of atomic structure based on quantum theory that was novel and admirable but could not easily have agreement with other physicists at that time … Without knowing why, he proposed two basic hypotheses. His first hypothesis was that [an] electron in [a] hydrogen atom while in circular motion in one of [the] orbits does not emit energy. The second hypothesis Bohr suggested was that energy was emitted when [an] electron is moving from one stable orbit to another. (Cha 2007, pp. 310–312; reproduced in Niaz et al. 2013)

Similarly, Cooper (1970) in his general physics textbooks recounts the difficulties involved with Bohr's model of the atom and that the mélange was not consistent (p. 325).

In order to pursue this aspect of Rutherford–Bohr models and the ensuing controversy further, Niaz et al. (2002) conducted a study within a history and philosophy of science framework in which general chemistry students were asked the following question: If Bohr changed Rutherford's model of the atom, in your opinion did Rutherford make mistakes while doing his experiments? These students formed part of an experimental group in the study in which they were encouraged to argue, discuss, and resolve contradictions. Interestingly, 54 % of the students responded correctly (in a written exam after the argumentation phase) by pointing out that Bohr's model improved with respect to Rutherford and this is how science progresses. Here are three examples of the reasoning provided by this group of students: a) Rutherford contributed by innovating and postulating a model that took us beyond that of Thomson … then thanks to the creative genius of Bohr who detected certain structural anomalies in Rutherford's model of the atom, we have progressed towards the true [?] structure of the atomic model (see Giere 2006a, who refers to such views as 'objectivist realism', also see Chap. 2); b) Rutherford's experiments constituted a base, which was later used by Bohr to establish his model … scientific process is characterized by the perfection of established theories … thus Rutherford's experiments were the points of departure for new discoveries; c) The model proposed by Rutherford constituted an extraordinary scientific advance for that time … but just as all models are subject to further tests by different scientists, Rutherford's model was changed by Bohr. (Niaz et al. 2002, p. 519)

For anybody familiar with general chemistry courses, the students' references to the following aspects of the scientific enterprise is surprising and encouraging indeed; the attempt to establish sequence of atomic models, Thomson, Rutherford, and Bohr (which was not required by the question), the perfection of established theories, the points of departure for new discoveries, and the models are subject to further tests by different scientists. By any standard, these responses indicate that given the opportunity within a domain-specific context of a chemistry course, students do understand the dynamics of scientific progress. Nevertheless, the reference to "true structure" is an indication that more work needs to be done. Furthermore, in this study 14 % of the students accepted that Rutherford made mistakes while doing his experiments.

Conclusion

Authors of most general chemistry textbooks analyzed in this chapter do understand and emphasize the role played by the visionary beliefs, philosophical arguments, hypothetical questions, and speculations of Democritus and some other Greek philosophers. However, after having recognized this, these textbooks go on to establish a difference between the Greek philosophers and Dalton, Thomson, Rutherford,

Bohr, and Sommerfeld models of the atom, which were based on reproducible experimental evidence. Consequently, in contrast to Chalmers (1998, 2009), who considered the philosophical ideas of the Greeks as an obstacle, these textbooks illustrate the changing nature of atomic theory by tracing the origin of atomic ideas to the Greeks. This clearly provides students an appreciation and understanding of the tentative nature of scientific knowledge, an important aspect of the nature of science.

Furthermore, in order to understand the change and transition from Greeks to Dalton to modern atomic theory (starting in 1897), it is essential to understand that the law of equivalent proportions (among other sources) provided the empirical evidence for Dalton's atomic theory (Rocke 2013c. Without recognizing Dalton's atomism as a theory, it is difficult to conceptualize the changing nature of atomic models. Of course, just like Thomson's, Rutherford's, Bohr's, and Sommerfeld's models of the atom were revised, changed, and improved, so was Dalton's atomic model.

Most textbooks analyzed in this chapter have an ahistoric presentation of the development of models of atomic structure. A historical reconstruction of the development of atomic models from Dalton to Bohr and beyond can provide students with an understanding of the dynamics of scientific progress and that it is difficult to interpret experimental data. On the contrary, most textbooks provide a simplistic presentation of scientific models and how these change with no reference to the difficulties and the controversies involved. For most textbooks, the inductive nature of science is paramount. Here are some examples of such presentations:

1. "What *ultimately banished this model to trash heap* was an experiment performed just a few years later, in 1909, by the British physicist Ernest Rutherford" (Russo and Silver 2002, p. 81, italics added).
2. "[Thomson's] model was *overthrown* in 1908 by another experimental observation" (Atkins and Jones 2008, p. 3, italics added).
3. "In 1908, this 'plum-pudding' model was *overthrown* by a simple experiment" (Jones and Atkins 2000, p. 7, italics added).
4. "Bohr approach was doomed to failure" (Whitten et al. 2013, p. 144).
5. "Niels Bohr had tied the unseen (the atom) to the seen (the observable lines of the hydrogen emission spectrum)—a fantastic achievement!" (Moore et al. 2011, p. 284).

Needless to say, history of science shows that models are not banished, overthrown, or doomed to failure but rather provide scientists with an opportunity to change, revise, and improve them. Despite some very good presentations of the historical context in some textbooks, most still follow the empiricist epistemology and hence make it difficult for students to understand the changing nature of atomic models. What is even more surprising and perhaps a cause for concern is the degree to which textbooks published in different countries and cultures follow the inductivist/empiricist epistemology (see Table 4.2 for a comparison of the textbooks published in the USA, Venezuela, Turkey, and Korea). This is precisely what Holton (1969) has referred to as the "experimenticist fallacy." Given the widespread nature of this orientation of the textbooks, collaboration between colleagues in different countries would help.

Although textbooks analyzed in this chapter overall do not emphasize the historical context of the changing nature of atomic models (quantitative scores are fairly low; especially see Tables 4.2 and 4.3), some textbooks do provide sufficient historical details (beyond that of simple mentioning the scientist's name) of the origin of chemical concepts and different aspects of the nature of science. The following are some examples of these textbooks that can be of help to students, teachers, and future textbook authors:

(a) Tentative or changing nature of atomic models: Atkins and Jones (2002), Denniston et al. (2011), and Seager and Slabaugh (2013)
(b) Greek philosophers: Bettelheim et al. (2012), Brown et al. (2012), Dickson (2000), Malone (2001), McMurry and Fay (2001), Raymond (2010), Russo and Silver (2002), Silberberg (2000), Timberlake (2010), Tro (2008), and Whitten et al. (2013)
(c) Dalton's atomic theory: Atkins and Jones (2002, 2008), Bettelheim et al. (2012), Brady et al. (2000), Brown et al. (2012), Chang (2010), Dickson (2000), Ebbing and Gammon (2012), Goldberg (2001), Hill and Petrucci (1999), Jones and Atkins (2000), Malone (2001), McMurry and Fay (2001), McQuarrie et al. (2011), Moore et al. (2002), Moore et al. (2011), Oxtoby et al. (2012), Raymond (2010), Russo and Silver (2002), Silberberg (2000), Timberlake (2010), Tro (2008), Umland and Bellama (1999), Whitten et al. (2013), and Zumdahl and Zumdahl (2014)
(d) Gay-Lussac's, Avogadro's, and Cannizzaro's contributions: Atkins and Jones (2008), Brady et al. (2000), Chang (2010), Dickson (2000), Ebbing and Gammon (2012), Goldberg (2001), Hill and Petrucci (1999), Jones and Atkins (2000), McMurry et al. (2007), McMurry and Fay (2001), McQuarrie et al. (2011), Moore et al. (2002), Moore et al. (2011), Oxtoby et al. (2012), Silberberg (2000), Spencer et al. (2012), Umland and Bellama (1999), and Zumdahl and Zumdahl (2014)
(e) Thomson's model: Brown et al. (2012), Hill and Petrucci (1999), Oxtoby et al. (2012), Silberberg (2000), and Whitten et al. (2013)
(f) Rutherford's model: Atkins and Jones (2002, 2008), Brown et al. (1997), Brown et al. (2012), Ebbing (1996), Ebbing and Gammon (2012), Hill and Petrucci (1999), Jones and Atkins (2000), Malone (2001), McMurry and Fay (2001), Oxtoby et al. (2012), Russo and Silver (2002), Silberberg (2000), Spencer et al. (2012), Timberlake (2010), Tro (2008), Umland and Bellama (1999), Whitten et al. (2013), and Zumdahl and Zumdahl (2014)
(g) Bohr's model: Chang (2010), Ebbing (1996), Ebbing and Gammon (2012), Malone (2001), Oxtoby et al. (2012), Silberberg (2000), Tro (2008), Umland and Bellama (1999), Whitten et al. (2013), and Zumdahl and Zumdahl (2014)

Finally, in order to respond as to why atomic models change and recapitulate the evidence provided in this chapter, let us consider the following presentations from two general chemistry textbooks:

> The theory of atomic structure has progressed rapidly, from a very primitive level to its present point of sophistication, in a relatively short time. Before we proceed, let us insert a note of caution. We must not think of the present picture of the atom as final. Scientific inquiry continues, and we should view the present theory as a step in an evolutionary process. *Theories are subject to constant refinement* …. (Denniston et al. 2011, p. 52, italics in the original)

> Our present understanding of the nature of matter is a model that has been developed and refined over many years. Based on careful observations and measurements of the properties of matter, the model is still being modified as more is learned. (Seager and Slabaugh 2013, p. 6)

In my opinion, most historians, philosophers of science, and researchers in science education would agree with the views expressed by these two general chemistry textbook authors. Interestingly, Denniston et al. (2011) have even inserted a note of caution as many students, and perhaps even teachers, may believe that we already have the final picture of atomic structure. Similarly, the reference to an "evolutionary process" is important if we want our students to understand that scientific inquiry continues and they too can still contribute toward its progress. To conclude, our understanding of atomic models is a never-ending quest that requires imagination, creativity, and innovative techniques in the laboratory to enhance our understanding. In this context, the tentative nature of scientific knowledge is perhaps the best theoretical framework for all introductory chemistry and science courses.

Chapter 5
Understanding Stoichiometry: Do Scientific Laws Help in Learning Science?

Introduction

Research in science education shows that both high school and undergraduate students have considerable difficulty in understanding stoichiometry (Agung and Schwartz 2007; BouJaoude and Barakat 2003; Dahsah and Coll 2007; Gabel 1993; Gabel and Bunce 1994; Garritz et al. 2013; Gultepe, et al. 2013; Niaz 1995a, 2008; Padilla and Furio-Mas 2008; Schmidt 1997; Staver and Lumpe 1995). Besides other factors, stoichiometry is a difficult topic as it also requires a conceptual understanding of several other concepts, such as the particulate nature of matter, the concept of mole, Avogadro's number, the conservation of matter, the balancing of chemical equations, and the laws of definite and multiple proportions. Furthermore, most chemistry teachers and textbooks emphasize problem-solving techniques based on algorithms that require "plug-and-chug" strategies and little conceptual understanding (Cotes and Cotuá 2014; Gulacar et al. 2014; Niaz 1995b; Niaz and Robinson 1993; Nurrenbern and Pickering 1987; Sawrey 1990; Tsaparlis 1998). Zoller and colleagues have explained that algorithmic problems require lower-order cognitive skills, whereas conceptual problems need higher-order cognitive skills (Tsaparlis and Zoller 2003; Zoller et al. 2002; Zoller and Tsaparlis 1997).

Based on these considerations, the objectives of this chapter are to (a) develop a theoretical framework based on epistemology, history, and philosophy of science (HPS) and (b) facilitate high school (grade 10) students' understanding of stoichiometry, based on an HPS teaching strategy.

M. Niaz, *Chemistry Education and Contributions from History and Philosophy of Science*, Science: Philosophy, History and Education,
DOI 10.1007/978-3-319-26248-2_5

Theoretical Framework

This study is based on the following history and philosophy of science (HPS) perspective: (a) teaching experiments, (b) constructivism, (c) scientific laws as idealizations, and (d) laws of definite and multiple proportions in chemistry.

Teaching Experiments

A basic premise of "teaching experiments" is to generate conflicts and contradictions between what the student knows and what we expect her/him to learn (Cobb and Steffe 1983). The instructor's role in the teaching experiment is to ask questions relevant to the learner's experiences that lead them into situations in which they experience conflicts or contradictions between their representations and those needed to interpret the new situations (Adey and Shayer 1994; D'Ambrosio and Campos 1992). Previous research has shown that such teaching experiments are quite helpful in facilitating conceptual change in chemistry (Niaz 2008).

This study does not assume that as the conflicting information is out there and the teacher or other students explained it to those who had difficulty with the problem situation, all would have seen the conflict. This line of argument misses the point that even if students can perceive the conflict, they resist changes in their preinstructional beliefs (cf. hard core of beliefs, Lakatos 1970). According to Chinn and Brewer (1993), "Instead of abandoning or modifying their preinstructional beliefs in the face of new conflicting data and ideas, students often staunchly maintain the old ideas and reject or distort the new ideas" (pp. 1–2). This clearly shows that the introduction of cognitive conflicts within teaching experiments requires considerable time, effort, and creativity on the part of both the students and the teachers.

Constructivism

According to Taylor (2015), "… constructivist theory is adaptable to many science teaching and learning scenarios, not in a simplistic sense as a method of teaching and learning but, …as a powerful epistemological 'referent' that enables teachers to think creatively about how to make learning science more motivating, memorable and meaningful …" (p. 223). Constructivism has been the subject of considerable research and controversy in the science education literature (Matthews 1997; Niaz 2011; Nola 1997; Osborne 1996). Furthermore, constructivism in science education has developed in many forms by drawing inspiration from various philosophical and epistemological sources. Of the different forms, radical, social, and psychological constructivism has enjoyed more popularity with science educators (Good 1993;

Perkins 2006; Phillips 1995). According to Kirschner, Sweller and Clark (2006) constructivist-based instructional approaches are not very effective due to recent research in the elaboration of the human cognitive architecture, namely a limited-capacity working memory which helps storage in long-term memory. Similarly, Tobias (2009) has concluded: "… there is stimulating rhetoric for the constructivist position, but relatively little research supporting it" (p. 346).

However, recent research has shown that training programs can improve students' cognitive capacities by leading to an increase in working memory (information processing) that can facilitate learning in science (Yuan et al. 2006). Similarly, St. Clair-Thompson et al. (2012) have reported that one of the best predictors of student performance on open-ended chemistry problems was mental capacity as determined by the figural intersection test based on Pascual-Leone's (1987) theory of constructive operators. These findings suggest that manipulation of the information processing load of chemistry problems can improve the performance of students who have limited mental capacity/working memory (for details, see Niaz 1988). This research shows that although constructivism is not a panacea, it can still facilitate important guidelines to the teacher provided that both "construction" and "explicit, guided instruction" knowledge are integrated (Duschl and Duncan 2009; Niaz 2001a). Furthermore, from a history and philosophy of science perspective, constructivism (similar to scientific theories) can benefit from continuous critical appraisals (Niaz et al. 2003). More recently, Niaz (2011) has shown that the contradictions faced by constructivism in science education can provide the rationale for its advance and evolution toward more progressive forms.

The dialectic approach to constructivism is an attempt to understand the reality within the context of complex interrelationships. According to Bidell (1988), "Rather than backgrounding conflict, the dialectic approach seeks to foreground it as the most salient feature of processes grasped in their complexity" (p. 332). The dialectic perspective emphasizes the understanding of psychological phenomena in their interrelationship to one another, rather than as isolated and separate processes as characterized by a Cartesian reductionist approach to science (cf. Lawler 1975; Pascual-Leone 1987; Piaget and Garcia 1989; Reese 1982; Riegel 1979). From a Piagetian (Piaget 1985) perspective, dialectics is closely linked to *équilibration majorante*, ". . . Piaget's all-encompassing constructive process for dealing with change; in particular with productive/ creative equilibration . . ." (Pascual-Leone 1987, p. 536). According to Riegel (1979), dialectic psychology incorporates divergent viewpoints and "… focuses on and tries to overcome the separation of organism and environment, consciousness and behavior, subject and object" (p. 27). Another important aspect of this study is the constructivist perspective, which ". . . presupposes that subjects construct their own world of experience (objects, events, transformations) by means of cognitive structures and organismic regulations/factors. This constructed world, however, is valid only if it epistemologically *reflects* distal objects, distal events and transformations actually occurring in the environment" (Pascual-Leone 1987, p. 534). Pascual-Leone (1976) goes beyond Piaget by postulating dialectical constructivism in which constructive theory attempts to ". . . model or reflect the subject's internal functional organization (i.e., his psychological system) in order to rationally reconstruct the genesis of the subject's performance" (pp. 90–91), similar to what Lakatos (1970, 1971,) referred to as the rational reconstruction of scientific theories and models in the history of science.

Niaz (1995b) has provided an illustration of how cognitive conflicts can lead students to construct models that represent progressive transitions and thus facilitate conceptual understanding. A cognitive conflict can be produced by various situations: (a) surprise produced by a result which contradicts a student's expectations, resulting in the generation of perturbations; (b) experience of puzzlement, a feeling of uneasiness, or a simple intellectual curiosity; and (c) experience of a cognitive gap, as if something in the student's knowledge structure was missing. Where cognitive conflict arises, and how students locate their resources (i.e., interact with those resources) in order to resolve the conflict, has been the subject of recent research in science education (Lee and Yi 2013). With respect to constructivism, Niaz (2011) has presented a framework based on a philosophy of science perspective that facilitates greater understanding.

Scientific Laws as Idealizations

In order to understand the nature of scientific laws, let us consider Newton's law of gravitation. According to Lakatos (1970), it is one of the "… best-corroborated scientific theory of all times …" (p. 92). Note that Lakatos refers to it as a theory. Feynman (1967) endorses the view that it is "… the greatest generalization achieved by the human mind" (p. 14). In spite of such impressive credentials, Cartwright (1983) asks, "Does this law (gravitation) truly describe how bodies behave?" (p. 57) and responds laconically, "Assuredly not" (p. 57). She explains further: "For bodies which are both massive and charged, the law of universal gravitation and Coulomb's law (the law that gives the force between two charges) interact to determine the final force. But neither law by itself truly describes how the bodies behave. No charged objects will behave just as the law of universal gravitation says; and any massive objects will constitute a counterexample to Coulomb's law. *These two laws are not true: worse they are not even approximately true*" (p. 57, emphasis added). The crux of the issue is that following Galileo's method of idealization (considered to be at the heart of all modern physics, by Cartwright 1989, p. 188), scientific laws, being epistemological constructions, do not describe the behavior of actual bodies. Newton's laws, gas laws, Piaget's epistemic subject—they all describe the behavior of ideal bodies that are abstractions from the evidence of experience, and the laws are true only when a considerable number of disturbing factors, itemized in the *ceteris paribus* clauses, are eliminated (Cartwright 1999; Kitchener 1993; Matthews 1987; McMullin 1985; Niaz 2009).

Chemistry students and teachers generally tend to understand the difference between scientific theories and laws in the following terms: a scientific theory has not been proved in its totality, whereas a scientific law has not only been proved but is also universal (Blanco and Niaz 1997, 1998). Ryan and Aikenhead (1992) reported that most students expressed a simplistic hierarchical relationship in which hypotheses become theories and theories become laws, depending on the amount of "proof behind the idea" (Lombardi and Labarca 2007). With respect to teachers, Smith and Scharmann (1999) reported, "Research over the past 45 years, however, has consistently shown that many American science teachers have a grossly inadequate understanding of the nature of science …" (p. 506). In contrast to these findings, Lakatos (1970), for instance, conceptualizes progress in science within a pluralistic model, in which "… the clash is not 'between theories and facts' but between two high-level theories: between an *interpretative theory* to provide the facts and an *explanatory theory* to explain them; and the interpretative theory may be on quite as high a level as the explanatory theory" (p. 129, italics in the original). This suggests that progress in science need not be characterized as a dichotomy between theories and laws, but rather as a "progressive problemshift" (Lakatos 1970), from one tentative theory to another.

Laws of Definite and Multiple Proportions

The framework presented in this section is helpful in understanding the laws of definite and multiple proportions within a history and philosophy of science perspective. Christie (1994) has traced the historical origin of the laws of definite and multiple proportions and presented an interpretation based on a history of science perspective. Various developments in chemistry have presented considerable problems for the law of definite proportions. In the case of the nonstoichiometric compounds (e.g., aluminum oxide), known as the "network solids," atoms are not bonded in discrete clusters as molecules but each to several neighbors in the form of a network. Similarly, synthetic polymers like nylon and polystyrene consist of large numbers of repetitions of a basic structural unit. According to Christie (1994), given these difficulties, the law of definite proportions, although is still mentioned in the textbooks, is not used in an explicit sense in modern chemistry (p. 616). In the case of the law of multiple proportions, the problem lies with the word "simple" or "small," which appears in its statement (Christie 1994). Although most of the time the ratios are small, there are thousands of different compounds containing just carbon and hydrogen, where the law is not instanced. It appears that "The law of definite proportions can be seen … as an exact rule with exceptions [and] the law of multiple proportions is not even a precise proposition" (Christie 1994, p. 619). In conclusion, Christie's work shows how even one of the most cherished laws in most chemistry textbooks, viz., the law of multiple proportions, is not even a precise proposition and concludes "… on a more revolutionary note,… many quite respectable laws of science are non-universal, and even that there are a few that cannot be formulated as precise propositions" (p. 613).

In this context, it is important to note that Giere (1995a, b) has presented an alternative account which provides a way of understanding the practice of science without the laws of nature in the following terms:

> But one need not appeal to history to deconstruct the concept of a law of nature. The concept is theoretically suspect as well. For example, any law of nature refers to only a few physical quantities. Yet nature contains many quantities which often interact one with another, and there are few if any truly isolated systems. So there cannot be many systems in the real world that exactly satisfy any purported law of nature. Consequently, understood as general claims about the world, most purported laws of nature are in fact false. So we need a portrait of science that captures our everyday understanding of success without invoking laws of nature understood as true, universal generalizations. (Giere 1995a, p. 10)

Although Giere espouses a naturalist philosophy of science, it is interesting to observe that there are many common elements in the treatments of Cartwright (1983, 1989), Christie (1994), Giere (1995a, b, 1999, 2006a, b), and Lakatos (1970) with respect to their understanding of laws and theories. All of them would subscribe to the thesis that scientific knowledge is tentative (except perhaps for Cartwright) and that it is advisable not to establish a dichotomous/hierarchical relationship between laws and theories. In other words, our knowledge progresses from one idea/hypothesis/theory to another, which is not ahistoric (Justi and Gilbert 1999). With this background, it is essential that science teachers reconsider the dichotomous presentation found in most textbooks of scientific progress in terms of theories and laws and that many of our well-known laws are in a sense "irrelevant" (cf. Blanco and Niaz 1997, 1998). The study reported in the next section is based partially on results reported by Niaz and Montes (2012).

Method

This study is based on two intact classes of high school students in Venezuela (grade 10, 15–16-year-olds). One class was designated as the control group ($n=32$) and the other as the experimental group ($n=31$). The two classes belonged to different schools and were taught by two different instructors. The instructor of the control group had a "Licenciatura" degree in chemistry education, had 25 years of experience, and used a traditional teaching strategy. The instructor of the experimental group (coauthor of this study, Niaz and Montes 2012) had a master's degree in chemistry education, had 23 years of experience, and used a constructivist strategy.

Traditional Teaching Strategy

The instructor of the control group used the traditional strategy in which laws of definite and multiple proportions are defined as definitive and irrefutable and are applied in the classroom as algorithms. This strategy can be summarized through the following sequence of steps: (a) definition of the two laws, (b) resolution of problems to illustrate the validity of the laws, and (c) resolution of additional problems based on algorithms.

Constructivist Teaching Strategy

The instructor of the experimental group used a constructivist teaching strategy (teaching experiment) based primarily on the theoretical framework presented above. The basic idea behind this strategy was the presentation of hypothetical experimental data leading to cognitive conflicts and a critical confrontation of different propositions, quite similar to what scientists do in order to achieve consensus. In order to facilitate understanding, the instructor used a modified form of the *learning cycle* (Lawson, Abraham, and Renner 1989). The learning cycle facilitates disequilibrium and argumentation and finally improves reasoning and is generally based on the following phases: (a) exploration, (b) term introduction, and (c) concept application. In this study, the instructor found the exploration and the concept application phases to be more useful.

Based on the history and philosophy of science framework (Cartwright, Giere, Lakatos), the instructor avoided defining the laws of definite and multiple proportions, unless the students themselves used these terms. Furthermore, due to classroom dynamics in this study, based on arguments, counterarguments, and conflicting propositions, it was not possible to follow the sequence of phases as suggested in the original learning cycle. The experimental teaching strategy lasted for 3 weeks, and in every session students were asked to present their ideas, arguments, and hypotheses, which were then confronted with counterarguments leading to critical discussions. In each of the problems, an attempt was made to achieve a consensus view with respect to the arguments that explained the problem situation well. For most students, this was a novel way of instruction in which they were not asked to solve algorithmic problems in order to have a right or wrong answer, but rather the instructor evaluated the strength and consistency of the arguments. Besides these considerations, the instructor of the experimental group followed the guidelines presented in the next section.

Guidelines for Facilitating Conceptual Change Based on a Dialectic Constructivist Strategy

Based on the theoretical framework of this study and on previous research (Niaz 2001b, 2008), it is suggested that the following guidelines can be helpful in facilitating conceptual change:

1. Examples to suggest that atoms are the building blocks of all matter. A brief mention of the Greek philosophers can be helpful.

2. Postulates of Dalton's atomic theory: presenting it as a research program can be helpful, which provides an example for the students, as to how a theory can be postulated without having all the experimental details worked out.
3. Modern atomic theory: electrons, protons, neutrons, and atomic masses.
4. Avogadro's number and the mole: discussion and resolution of problems.
5. Discussion and significance of molecular formulas. Students were presented with a table in which the chemical composition of various samples of the same compounds was included. Students were asked the following question: do the different samples represent the same compound? In the discussion that followed, students were asked to justify their response. It was hypothesized that after the discussion, without enunciating the law of definite proportions, students would understand that "The great majority of compounds have a fixed and definite atomic composition; we call such compounds *stoichiometric*. All gaseous compounds are stoichiometric and are composed of discrete (individual) molecules or atoms. The formula of a substance specifies its atomic composition" (p. 30, Segal 1989, italics in the original). For example, carbon dioxide has the formula CO_2. This means that one molecule of carbon dioxide contains one atom of carbon and two atoms of oxygen.
6. Dalton's atomic theory was the first to provide an explanation of the formation of stoichiometric compounds.
7. Describe how nitrogen and oxygen form various compounds with different molar formulae, for example, N_2O, NO, and NO_2. Once again, at this stage, Segal (1989), in contrast to most textbooks, does not enunciate the law of multiple proportions but provides the following description: "Note that nitrogen and oxygen combine in several different ratios, that each is ratio of small integers, and that each combination results in a different substance with different physical and chemical properties" (p. 31).
8. Based on his atomic theory, Dalton predicted and then explained the formation of more than one compound by the same two chemical elements. These experimental findings provided strong support for Dalton's atomic theory. This provides an opportunity to familiarize students with the complex relationship between theory and experiment that goes beyond the positivist presentations found in most textbooks.
9. Presentation and discussion of the research programs of Dalton and Gay-Lussac and how this led to a scientific controversy. This provided an opportunity to illustrate how competition between rival theories can provide a better explanation.
10. Emphasize molar ratios rather than weight (mass) relationships between different elements, which combine to form compounds. For example, Segal (1989) provides the following example: "The commonly used pain reliever, aspirin has the molecular [molar] formula $C_9H_8O_4$. If a sample of aspirin contains 0.968 g

of carbon, what is the mass of hydrogen in the sample?" (p. 33). In order to solve this problem, students are obliged to conceptualize the molar ratio of carbon to hydrogen of 9:8. At this stage, it is interesting to observe that the enunciation of the law of definite proportions leads to an emphasis on the macro aspects (percentage composition of the elements in a compound) and sort of ignores the micro aspects (particulate nature of matter). For the importance of particulate nature of matter in chemistry education research, see Chap. 1.

11. Balanced chemical equations. After explaining the difference between empirical and molar formulae, Segal (1989) goes on to describe the concept of a balanced chemical equation in the following terms: "When we speak of a balanced chemical equation, we mean an equation that describes a physical or chemical change, and is consistent with the requirement that *in any process both mass and charge are conserved*, that is, they remain the same before and after the change has taken place" (p. 39, italics in the original). It is interesting to note that Segal (1989) does not enunciate the law of conservation of mass here or any place else in her textbook.

Classification of Students' Responses

The students in the experimental group participated in the constructivist teaching strategy for 3 weeks. The students in the control group also participated in the traditional strategy for the same period of time. After 2 weeks, both groups of students were evaluated on a semester exam which consisted of six items (two algorithmic and four conceptual; results from only three conceptual items are reported here, cf. Niaz and Montes 2012 for complete details). The second author of this study was present during evaluation of the control and experimental groups. Items 1–3 were conceptual problems requiring argumentation, restructuring, and reasoning based on alternative interpretations. Item 1 was adapted, with small modifications, from Rodgers (1995, p. 216). Items 2 and 3 were adapted, with some modifications, from Christie (1994) and Niaz (2001b).

For the conceptual problems (Items 1–3), students responses were classified as (a) conceptual, if a student explicitly elaborated arguments in a coherent and logical manner from the data given in the problem; (b) partially conceptual, if a student attempted to present some arguments without a logical and coherent scheme; and (c) rhetorical, if a student reproduced some elements of the problem situation with no attempt to present arguments in a consistent fashion. (For further details, see Niaz and Montes 2012.) Students' responses were translated from Spanish into English, first by both authors separately. In the case of differences, these were translated again and compared until consensus was achieved.

Results and Discussion

How to Explain a Contradiction and the Ensuing Cognitive Conflict? (Based on Students' Responses on Item 1)

Item 1
Iron (Fe) and oxygen (O_2) form two oxides. One of the oxides having Fe (II) has the formula FeO. This shows that one atom of Fe will combine with one atom of oxygen. In other words, if 100 g of iron combine with 28.65 g of oxygen, we would obtain 128.65 g of FeO. However, it has been found that only 95.00 g of Fe combines with 28.65 g of oxygen. *Frequently, advances in scientific knowledge are based on such contradictions.* How can you explain this contradiction?

This is a conceptual problem as the response does not depend on memorization of formulae or algorithmic procedures but rather on argumentation, restructuring, and the capacity to develop alternative models/hypotheses, in the face of conflicting information. Furthermore, students in this study were not previously exposed to such problems. The problem presents to students a situation in which the same compound, iron oxide (Fe^{2+}), has a different composition and they are asked to explain this contradiction. Most students found this problem to be difficult, as the problem required them to think, go beyond pre-elaborated responses, resolve a conflict (Lee and Yi 2013), and in a sense overcome the myth of the correct response. Only three (10 %) students from the experimental group and none from the control group responded conceptually. Here are two examples of conceptual responses provided by the experimental group:

> This problem reports the finding of the same compound with the two elements combining in different proportions. In my opinion, changes must be introduced in the law of definite proportions. In other words, it must be explained as to why the same compound exists with different proportions and when/which are the exceptions to this law (Student #1).

> We must avoid being carried away by generalities as they are not complied in all cases. One of these cases is that of iron (II) oxide, in which the two elements can form distinct types of iron oxide. The conditions in which a compound is formed can modify the mass relationship of the combining elements, thus forming different compounds from the same elements (Student #12).

The response of Student #1 indicates acceptance of the law of definite proportions and at the same time points out that it is not enough to explain the contradiction presented in the problem. This recognition can be considered as the need for alternative explanations. It is interesting to note that on first reading of the exam question (Item 1), this student stated: "This question refers to the law of definite proportions, and laws cannot be contradicted. A law is a law and this is not open to discussion" (reproduced from the second author's class notes. Students in this study were supposed to read and respond in silence, as in any evaluation. Perhaps, the novelty of the question led this student to express himself loudly, which is very unusual. This was the only student who made such a comment). This observation is important, because during the experimental strategy the laws of definite and multiple proportions were not mentioned. However, it is plausible to suggest that interactions with students from other classes and textbooks provided students with the necessary information with respect to these laws. With this perspective, it is important to note that Student #1, based on the contradiction, changed his perspective with respect to what he expressed initially and thus resolved the ensuing cognitive conflict. Student #12 refers to the need to go beyond generalities as under different conditions; the same elements can form compounds with different compositions. Such changes implicitly recognize the need for better explanations that go beyond the accepted laws and are a manifestation of the tentative nature of scientific theories (Lakatos 1970) and approximates to what Giere (1999) has referred to as "science without laws." These changes can be attributed to the intervention in which students were constantly asked and encouraged to provide alternative explanations in the face of contradictions.

Six (19 %) students from the experimental group and none from the control group provided a partially conceptual response to Item 1. Here are two examples from the experimental group:

> … although elements combine in a definite relation with respect to their masses, I think this depends on the type of elements that react—because there are elements that are in excess and others that are limiting reagents … (Student #31).

> The relation in which these elements combine in the two cases are different. This means that these are different compounds with the same properties (Student #33).

Responses in this category, on the one hand, show a tendency to dissociate from the idea of chemical combination as definitive and constant but on the other hand do not argue convincingly to understand that the same compound can have different compositions.

Eighteen (58 %) students of the experimental group and 29 (91 %) of the control group responded with a rhetorical response, and the difference is statistically significant (chi-square, $p < 0.05$). It is plausible to suggest that the higher percentage of rhetorical responses by the control group is an indicator of lack of understanding of the problem situation. The following are two examples of rhetorical responses by experimental group students:

> In my opinion, as we are combining 100 g of iron + 28.63 g of oxygen = 128.63 g of FeO. Thus it can be concluded that this compound does not need additional Fe in order to combine with oxygen to form FeO (Student #2).

> Perhaps atmospheric conditions were responsible for having left a certain amount of iron without reaction or some iron dispersed into air due to disintegration (Student #3).

> Iron is the reagent in excess and oxygen the limiting reagent (Student #10).

Despite the difficulties (cognitive conflicts, Chinn and Brewer 1993) involved in responding to Item 1, it is suggested that at least some students understood that contradictions are part of the scientific endeavor and that they lead to alternative interpretations and explanations. This understanding is important if we want our students to understand that progress in science is intricately associated with the tentative nature of science.

How to Draw Conclusions Consistent with the Data? (Based on Students' Responses on Item 2)

Item 2

Copper (Cu) forms two oxides. 100.00 g of copper combines with 12.598 g of oxygen to produce Cu_2O or with 25.196 g of oxygen to produce CuO. The amount of oxygen that combined in CuO is double that of Cu_2O. In other words, the relation of the mass of oxygen that combined in the two compounds is 12.598:25.196 or simply 1:2. According to these results:

(a) Can we conclude that if two elements combine to form more than one compound, then the different masses of one of these elements that combine with a fixed mass of the other would do so in small whole-number ratios?
(b) Is this conclusion consistent with results presented here?

Justify your response.

This item deals explicitly with the law of multiple proportions. Students in the control group were given the definition of the law and solved various problems, whereas students in the experimental group were not given the definition but only solved problems dealing with such situations. It is possible that students in the experimental group may have been aware of the algorithmic form of the law through interactions with students of other classes or through their textbooks. Six (19 %) students of the experimental group and two (6 %) of the control group responded in conceptual terms. Here are three examples of conceptual responses from the experimental group:

> Two atoms can combine to form various compounds and one of the two elements would not change its mass and combine with the other in small whole-number ratios. However, it is possible that this relation may exist as a big whole-number ratio (Student #1).

> Yes, two elements can combine to form different compounds, so that if the proportion in mass of one is constant, that of the other varies. Data presented is consistent with the conclusion (Student #12).

> The mass relation in which oxygen combines is 1:2, which is small and the conclusion is consistent with the data presented (Student #22).

The response of Student #1 is interesting, as it considers the possibility of whole-number ratios in which elements combine, to be large. Interestingly, this is what happens in Item 1. It is possible that this student may have solved Item 1 before Item 2.

Eight (26 %) students from the experimental group and only one (3 %) from the control group provided a partially conceptual response, and the difference is statistically significant (chi-square, $p<0.05$). Here are two examples of partially conceptual responses from the experimental group:

> At times the whole-number ratio is small. It all depends on the mass of the compound (Student #17).

> I think that the conclusion is consistent with the results, as the example is quite consistent. Copper combines with oxygen in small whole-number ratios and with a fixed mass (Student #21).

Ten (32 %) students from the experimental group and 29 (91 %) from the control group provided rhetorical (see "Method" section) responses, and the difference is statistically significant (chi-square, $p<0.001$). It is plausible to suggest that students in the control group had greater difficulty in understanding this problem and this is why the percentage of rhetorical response for this group is higher. This clearly shows the importance of the intervention. Here are some examples from the experimental group:

> According to the results, in order to form two compounds the mass of one must change (Student #4).
>
> In this case, as oxygen has two atoms its quantity is less. However, in the case of one atom the quantity is greater (Student #20).
>
> I am in agreement with the conclusion (Student #32).

These responses show that this group of students either did not understand the problem situation or simply could not process the relevant information and hence resorted to a simply rhetorical strategy. It is plausible to suggest that given the difficulty of such problems, perhaps these students needed a more extended teaching intervention.

When Is It Justified to Generalize? (Based on Students' Responses on Item 3)

Item 3

Consider the following compounds formed by carbon (C) and hydrogen (H):

(a) Ethyne (C_2H_2) and ethylene (C_2H_4)
(b) Butane (C_4H_{10}) and heptane (C_7H_{16})
(c) Ethyne (C_2H_2) and pentane (C_5H_{12})

(i) Do you think that the three cases mentioned here can be considered as instances of the generalization presented in Item 2?
(ii) Do you think we need an alternative explanation?

Justify your response.

This was a difficult and thought-provoking question that could be considered as a follow-up to Item 2, in which the formation of Cu_2O and CuO was presented as an instance of the law of multiple proportions (only the students in the control group were provided with the definition). In contrast, Item 3 provided students with three examples of fairly well-known hydrocarbons, in which only the first case (C_2H_2 and C_2H_4) could be considered as an instance of the law of multiple proportions. This item provided an opportunity to observe the difference in the problem-solving strategies of the two groups of students, viz., one that was exposed to an algorithmic definition of the law (control group) and the other that was provided an opportunity to think and reason (experimental group). Only one (3 %) student from the experimental group and none from the control group provided a conceptual response. Here is the conceptual response of the student from the experimental group:

> The generalization is valid for case (a), as the mass of hydrogen in C_2H_4 is double that of C_2H_2. In cases (b) and (c), it all depends on what we may consider as the fixed mass. However, it is not appropriate to generalize in these cases. Each one of such cases should be analyzed and studied in order to find alternative explanations, not only for this problem but for others that we may find in the future (Student #3, before presenting this response, the student did all the relevant calculations for the three cases of hydrocarbons). Note: This student provided a partially conceptual response on Item 2.

Six (19 %) students from the experimental group and only one (3 %) from the control group provided a partially conceptual response. Here are some examples of responses from the experimental group:

> In each case there exists a different relation, as hydrogen in order to form different compounds varied its mass (Student #1, all relevant calculations were presented). Note: Apparently, this student did not follow-up on his/her response in Item 2, where he had suggested the possibility of big whole-number ratios.

> We can observe that in each one of the compounds, the proportion of the same combining elements is different. All these relations are in small whole-number ratios (Student #22, all relevant calculations were presented).

It is interesting to note that students in this group did all the relevant calculations and then concluded that all whole-number ratios were small (i.e., C_2H_2 and C_2H_4) and simply ignored the cases in which this was not the case (viz., C_4H_{10} and C_7H_{16} and C_2H_2 and C_5H_{12}). One student (#16), however, recognized that the generalization presented in Item 2 can perhaps only be applied in case (a).

Fourteen (45 %) students from the experimental group and 23 (72 %) from the control group provided rhetorical (see "Method" section) responses. Here are three examples of responses from the experimental group:

> Yes, because in the previous cases 2 atoms combined, and in the present case various atoms did so (Student #5, no calculations were performed).

> The generalization is valid for this problem, as different compounds are formed from the same elements. Thus it is not necessary to look for an alternative explanation (Student #12, all relevant calculations were presented).

> The generalization is valid because in each case there are two formulae of the same elements. The only difference is in the form of combination (Student #18, no calculations were presented).

In general, rhetorical responses of both experimental and control group students did not perform all the calculations and concluded that the generalization from Item 2 could be applied to all three cases presented in Item 3. Once again, the percentage of students from the control group who gave a rhetorical response is much higher than that of the experimental group. Results obtained also show that none of the students from either the control or the experimental group solved all three items (1, 2, 3) conceptually. Four students from the experimental group and none from the control group solved two items conceptually. Nine students from the experimental and four from the control group solved one item conceptually. Furthermore, students in the experimental group also performed better on the two algorithmic problems included in the study (for further details, see Niaz and Montes 2012).

At this point, it is constructive to report results from a previous study in which the same problem (Item 3) was presented to chemistry students who were about to finish their "Licenciatura" degree (a 5-year course with dissertation) in chemistry. Results showed that five of the seven students were reluctant to question the utility of the law (multiple proportions) in chemistry (for details, see Niaz 2001b). One of the students who questioned the law reasoned: "This shows that the law is a limited one and does not explain many cases. We can conclude that such laws are not absolute, but rather explain a phenomenon only under certain conditions" (reproduced in Niaz 2001b, p. 252). This also shows how students, like scientists, resist changes on issues that are considered to be fundamental and form part of the "hard core" of their epistemological beliefs (cf. Chinn and Brewer 1993; Lakatos 1970). Commenting on such student responses, Christie, a historian of science, has provided food for thought:

> … student responses illustrate to me a problem that I continually find with science students—an unfamiliarity with thinking critically; an expectation that any question has a 'right' answer, and a mechanical/algorithmic route to finding that answer. In my view, one of the most important contributions that a history and philosophy of science module can and should make to a science student's education is a challenge to break that mould. (reproduced in Niaz 2001b, p. 253)

In order to elicit alternative opinions, I sent a copy of my article (Niaz 2001b) to Alan Rocke, an eminent historian of chemistry. Rocke (2013a) responded by suggesting that the laws of simple and multiple proportions are nothing more than special cases of the law of equivalent proportions and that "I wonder if your students would have been more successful in seeing the positive instances of stoichiometry if you had asked them to find instances of the law of equivalent proportions, and not just constant or multiple proportions." On reading these comments, it occurred to me that the results reported in this chapter could also be interpreted within an alternative framework, and so I sent a preliminary version of this chapter with a brief summary and the following message to Alan Rocke: "It appears to me that the experimental group students, in a sense used the law of equivalent proportions, without of course having been explicitly instructed about it. I wonder, if you agree with my interpretation." Rocke (2013b) responded in the following terms: "I think that your suggestion that the experimental students were in effect rediscovering and making use of the law of equivalent proportions seems very plausible to me." This clearly shows that if the students are given an opportunity, they can look for alternative and creative ways of understanding data, instead of the traditional classroom in which laws are memorized in the form of formulae and applied without much learning.

Indeed, teaching chemistry within a history and philosophy of science framework can provide students an opportunity to not only appreciate how scientists work but also develop a more critical stance toward scientific development and progress (Niaz 2012a).

Conclusions and Educational Implications

In the study presented above, the students in the control group were exposed to traditional problem-solving strategies in which algorithms based on formulae were used. In contrast, students in the experimental group participated in argumentation, cognitive conflicts, and historical reconstruction of events, which facilitated them to develop alternative models/hypotheses in the face of conflicting information. This experimental teaching strategy was based on an epistemological, historical, and philosophical framework, which required considerable time and effort on behalf of both the students and the instructor (cf. Chinn and Brewer 1993; Lee and Yi 2013 for the difficulties involved).

Interestingly, results obtained in this study show the better performance of the experimental group not only on all conceptual items (Items 1, 2, 3) but also on the two algorithmic problems. Despite Christie's (1994) critique, textbooks still emphasize the law of definite proportions. One study has reported that of the 27 general chemistry textbooks analyzed (all published in the USA), only three explained chemical combination, stoichiometry, and other related concepts without enunciating or referring to the law (Niaz 2001b). This study shows that experimental group students were better prepared to understand and interpret the formation of nonstoichiometric compounds. Item 2 (dealing with the law of multiple proportions) is quite typical in most traditional classes, including the one (control group) in this study, and thus the control group had fair amount of experience (and advantage) in solving similar problems. This study shows that even on Item 2, students in the experimental group performed much better, thus providing the rationale and plausibility of implementing similar experimental strategies in high school and introductory university level courses (dialectic constructivist, HPS perspective). Once again, Christie (1994) has shown that even if one could consider the law of definite proportions as an exact rule with exceptions, in contrast the law of multiple proportions is not even a precise proposition. Interestingly, nine (out of 27 analyzed) general chemistry textbooks did not mention the law of multiple proportions (Niaz 2001b). Item 3 was the most difficult for students in this study and illustrates how the law of multiple proportions is not instanced by the thousands of compounds formed by carbon and hydrogen. This study shows that despite the difficulty of Item 3, students in the experimental group performed better than those in the control group.

Finally, based on the theoretical framework and the results obtained, this study has important educational implications. If scientific laws are idealizations, then they do not describe the behavior of actual bodies and hence may be of limited help in understanding the world of experience (Giere 1999). Furthermore, in his naturalistic philosophy of science, Giere (2006a) has further clarified this:

> It is first necessary to be clear that the notion of a law of nature is not part of the literal content of any science. It is a notion that belongs to a meta-level interpretation of what scientists do, for example, that they discover laws of nature. Thus, to question the applicability of this notion is not to question any science itself but, rather, only *how the aims and achievements of scientific activities are to be described.* (pp. 69–70, italics added)

Interestingly, this is what science education needs to do, namely, help students *understand the aims and achievements of scientific activities*, i.e., how scientists think, design experiments, interpret, and argue with their peers. This perspective leads to a critical evaluation of the laws of definite and multiple proportions and their role in chemistry education. Emphasizing these laws (instead of understanding) inevitably leads to memorization (the use of algorithms and formulae) in learning chemistry and especially stoichiometry. This study shows that stating the laws and their definitions first and then asking students to work out exercises as illustrations of the laws is not a very productive strategy (cf. Stinner 1992). Similarly, Lakatos was opposed to a strategy which suggests define your terms before you start teaching (cf. Brown 1990). Precisely, this study shows that it is the context of a problem that can lead students to understand how stoichiometric relations are established. Guidelines presented in this study, based on a dialectic constructivist strategy (HPS perspective), can facilitate students' conceptual understanding.

Chapter 6
Understanding Valence Bond and Molecular Orbital Models: Contingency at Work

Introduction

A historical study of the development of the model of the chemical bond shows that the idea of sharing electrons (covalent bond) posed considerable conceptual constraints to scientists (Kohler 1971). Therefore, it is no surprise that high school and university students consider the topic to be difficult. In an attempt to simplify the topic, most textbooks present rules (algorithms) for writing simple Lewis structures for covalent bonds, which are memorized by students. This chapter provides an alternative by reconstructing the historical development of the model of covalent bonding that leads to the incorporation of more recent models with greater explanatory power.

G.N. Lewis (1916) was perhaps the first chemist to introduce the idea of sharing electrons to form covalent bonds, which later led to understanding the shape of the molecules. According to Coffey (2008):

> There was a great deal of extraneous model building and speculation in Lewis's paper—the cubic atomic structure with static electrons stuck at the corners seems quaint today—but the lasting concept from the paper was what later became known as the *covalent bond,* the sharing of an electron pair between two atoms. (p. 138, original italics)

This clearly shows that the essence of Lewis's ideas (cubic atom, sharing of electrons) led to the formulation of model of the covalent bond. Starting in 1919, I. Langmuir started popularizing Lewis's idea by introducing what he called the "octet theory" (a configuration of eight electrons) and later coined the term "covalent bond." This led to a serious priority dispute between Lewis and Langmuir with respect to the origin of the model of the covalent bond. However, it is interesting to note that Arthur Lamb, then editor of *Journal of the American Chemical Society*, agreed with Lewis that Langmuir's octet theory in essence was based on Lewis's 1916 article.

M. Niaz, *Chemistry Education and Contributions from History and Philosophy of Science*, Science: Philosophy, History and Education,
DOI 10.1007/978-3-319-26248-2_6

Lewis's Postulation of the Covalent Bond

An important part of Lewis's postulation of the model of the covalent bond was the cubic atom, according to which electrons in an atom are arranged symmetrically at the eight corners of a cube. This later led to the formulation of the "rule of eight" or the "octet rule." Thus, the single bond was conceived of as two cubic atoms with a shared edge (pair of electrons) and the double bond as two cubes with a common face. Furthermore, Lewis introduced the idea of sharing electrons that at first appeared to be a bizarre idea, as due to Coulomb's law two electrons should exert a force of repulsion (Rodebush 1928). According to Kohler (1971):

> When it was first introduced, Lewis's theory was completely out of tune with established belief. For nearly 20 years it had been almost universally believed that all bonds were formed by the complete transfer of *one* electron from one atom to another. The paradigm was the ionic bond of Na^+Cl^-, and even the bonds in compounds such as methane or hydrogen were believed to be polar, despite their lack of polar properties. From the standpoint of the polar theory the idea that two negative electrons could attract each other or that two atoms could share electrons was absurd. (p. 344, original italics)

Lewis emphasized the need for chemists to master the laws of physics in order to understand the electron-pair bond, which he considered to be "the cardinal phenomenon of all chemistry." The introduction of quantum mechanics for understanding the covalent bond led to a conflict for many chemists, as it seemed to overlook and perhaps even threaten the chemists' concern for visualization versus thinking mathematically (cf. Niaz 2013).

A theoretical explanation for the sharing of electrons (before the concept of spin) was first provided by Pauli's exclusion principle (Pauli 1925), just as the cubic atom (as postulated by Lewis) did previously. Lewis explained how two unpaired electrons in different atoms might be coupled magnetically and form the nonpolar bond (cf. Rodebush 1928, for an early recognition of Lewis's contribution). More recently, Gavroglu and Simões (2012) have recognized the novelty of Lewis's electron-pair bond for understanding chemistry:

> Lewis's choice of representing the atom as a succession of concentric cubes [cubic atom] played a crucial role in the suggestion of the shared electron-pair bond. This novel idea, which grew out of the exploration of a pictorial representation in the context of suggestions by other scientists and Lewis's own musings over the matter, introduced a new theoretical entity—the shared electron pair into chemistry. (p. 53)

It is plausible to suggest that if the sharing of electrons was considered to be "bizarre" and "absurd" by scientists (Kohler 1971), it could appear counterintuitive to students as well (for further details, see Niaz 2001c). The controversial origin of the covalent bond model and its rivalry with the ionic bond model provides a good opportunity to illustrate how progress in science is based on controversy and that established theories or ways of thinking are difficult to change. Furthermore, Pauli's exclusion principle provides a better theoretical explanation for the sharing of electrons, just as Lewis's cubic atom did previously. The role of Pauli's exclusion principle for chemistry education was recognized early by Gillespie (1963), and it has been an integral part of later developments of the model of the covalent bond. The transition from Lewis's cubic atom to Pauli's exclusion principle to what came next provides an illustration of how scientific knowledge is tentative. According to Project 2061, "The notion that scientific knowledge is always subject to modification can be difficult for students to grasp. It seems to oppose the certainty and truth popularly accorded to science, and runs counter to the yearning for certainty that is characteristic of most cultures, perhaps especially so among youth" (AAAS 1993, p. 5). Indeed, this clearly shows the importance of this aspect of the nature of science discussed in Chap. 3. For a discussion of "truth in science" or "true theories," see Chap. 2 and also Chap. 8.

How the model of the covalent bond developed further is the subject of the next section. This also sets the stage for introducing the valence bond model, the molecular orbital model, and the notion of contingency and its role in scientific development. Discussing how these are presented in general chemistry textbooks is a major objective of this chapter.

Covalent Bond and Quantum Mechanics

The molecular shape and the geometry of chemistry molecules play an important part in determining chemical properties, such as reactivity, odor, taste, and drug action. Lewis structures (as presented in general chemistry textbooks and in the previous section) are based on a localized electron model and only serve to present the distribution of electron pairs in very simple molecules.

The first satisfactory model of the hydrogen molecule based on quantum mechanics was presented by Heitler and London (1927). This model specifically based bond stability on Pauli's exclusion principle. Next, the contributions of Slater and Pauling led to the Heitler–London–Slater–Pauling model, now known as the valence bond model. Pauling (1931) first explained the nature of the chemical bond based on quantum mechanics, which later formed part of his classic *The Nature of the Chemical Bond* (Pauling 1939). Resonance and hybridization were two of Pauling's major contributions toward the understanding of the chemical bond. Nevertheless, both resonance and hybridization raised many ontological issues and remained lifelong concerns for Pauling. A major concern of his theory was visualizability, which was of particular interest to chemists. Interestingly, however, the scientific community was aware of the pitfalls concerning both schemata: (a) were orbitals real? and (b) was resonance real? The debate continued even into the 1990s, when Ogilivie (1990), p. 285) pointed out that "... there are no such things as orbitals" and that hybridization could not explain the tetrahedral structure of methane. Pauling (1992) responded that hybridization and the tetrahedral structure were verified experimentally.

The valence-shell electron-pair repulsion (VSEPR) model proposed by Gillespie and Nyholm (1957) is generally considered to be a part of the valence bond model (Gillespie, 1963). Possible extensions of the VSEPR model are discussed in

Gillespie (2008). More recently, Gillespie (cf. Cardellini 2010, p. 483) has recognized the central role of Pauli's exclusion principle in the VSEPR model: "Electrons with the same spin have a zero probability of being found simultaneously at the same point in space, a low probability of being found close together, and are most probably to be found as far apart as possible."

Mulliken (1932) extended Bohr's and Pauli's *aufbau* principle to molecules based on quantum mechanics, which led to the formulation of molecular orbitals. Besides spectroscopic data, Pauli's exclusion principle played an important role in Mulliken's approach to chemical bonding, which later became the molecular orbital model. Mulliken argued that besides bonding and nonbonding electrons, there were those that opposed bonding, which he called *antibonding electrons*. This was considered to be a critique of the Slater–Pauling approach, starting a controversy that continued for many years.

Both Pauling and Mulliken applied quantum mechanics to the valence theory and developed the valence bond (VB) and molecular orbital (MO) models, respectively. Pauling presented his quantum mechanical treatment in a seminal book (Pauling and Wilson 1935). Interestingly, Pauling (1980, p. 39) also used the molecular orbital method to make quantum mechanical calculations for substitution in the benzene ring. However, he expressed his concern for general chemistry courses in cogent terms: "Only one system for treating valence, valence bonds, and molecular structure should be used for the elementary students, in order that he build up a sound *picture of molecules* and the chemical bond and not be confused" (Pauling 1980, p. 39; italics added). Actually, this was the major difference between the two models: VB emphasized visualization and thus continued the tradition of Lewis, whereas MO treatment was much more mathematical and departed from classical valence theory. According to Gavroglu and Simões (2012), "Pauling and Slater justified the visual models of traditional chemists through skillful application of the mathematical language of quantum mechanics" (p. 30). In science education, Gilbert (2005) and others have recognized the importance of not only models but also visualization in order to enable students "to think like a scientist." Similarly, Niaz (2010, 2012) has drawn attention to the need for teaching science as practiced by scientists (also see Chap. 2).

At this stage, it is important to note that although models are not subordinate to theory and data (Morrison and Morgan 1999, p. 36), they do "… teach us about both theories and the world by providing concrete information about real physical and economic systems" (p. 24). According to Giere (1999), "… we understand the word 'theory' as including both the cluster of models and a broad range of hypotheses utilizing these models" (pp. 167–168). Based on this relationship between theories and models, Giere (1999) then illustrates the role of the two-chain model of DNA presented by Watson and Crick:

> … mere agreement with the measured water content was not regarded as evidence for the two-chain model. Many different models can satisfy that demand. What saved the day in this case was the prior judgment that there are no plausible alternative models that predict the highly specific observed X-ray pattern … a good indicator that the proposed model *fits better* than any others regarded as plausible rival models. (p. 75, original italics)

The role of other researchers and alternative models in the determination of the DNA structure is recognized even by a general chemistry textbook:

> The model of DNA established by Watson and Crick was based on key contributions of other researchers, including the analysis of the base composition of DNA. The analysis of DNA from many different forms of life revealed an interesting pattern. The relative amount of each base often varied from one organism to another, but in all DNA the percentages of adenine and thymine were always equal to each other as were the percentages of guanine and cytosine. (Seager and Slabaugh 2013, p. 641)

Consequently, it is plausible to suggest that both the VB and the MO are rival models (similar to theories) that can and do compete with each other. Interestingly, the degree to which VB and MO models predict observed molecular shapes varies.

Coulson (1937), originally trained in physics and mathematics, also made important contributions to the development of the molecular orbital model. He was particularly aware of chemists' need to apply quantum mechanics in a way that retained the pictorial aspects of visualizing molecules. The publication of *Valence* by Coulson (1952a) immediately became a success and a strong competitor to Pauling's *The Nature of the Chemical Bond.* Coulson was particularly aware of Pauling's views and took care to provide a fair treatment to both the molecular orbital and the valence bond models.

The Contingency Thesis in the History and Philosophy of Science

How science is practiced by scientists, namely, its contingent nature based on creativity and insight, has been recognized by physicist–philosopher James Cushing (1989) in cogent terms: "Science is an historical entity whose practice, methods and goals are *contingent*. There may not be *a* rationality which is the hallmark or the essence of science" (p. 2, italics in the original). In a footnote, Cushing explains what he means by *contingent*: "I simply mean not fixed by logic or necessity" (p. 20). Cushing (1994) has explained the role played by the contingency thesis in the development of quantum mechanics:

> The central theme of this book is that historical contingency plays an essential and ineliminable role in the construction and selection of a successful scientific theory from among its observationally equivalent and unrefuted competitors. I argue that historical contingency, in the sense of the *order* in which events take place, can be an essential factor in determining which of two empirically adequate and fruitful, but observationally equivalent, scientific theories is accepted by the scientific community. (Preface, p. xi; italics in the original)

As an example of historical contingency, Cushing (1998) considered that around 1927, besides the Copenhagen interpretation of quantum mechanics, there were two rival interpretations, namely, Schrödinger's wave picture and de Broglie's pilot wave model (a precursor to Bohm's theory of hidden variables, cf. Cushing 1994). In order to respond to the question what happens to the rival interpretations, Cushing

(1998, p. 353) has provided a plausible albeit picturesque answer: "Copenhagen got to the top of the hill first and, to most practicing scientists, there seems to be no point in dislodging it." In retrospect, it is interesting to note that in a critical review a physicist has conceded that "At the turn of the century, it is probably fair to say that we are no longer sure that the Copenhagen interpretation is the only possible consistent attitude for physicists … Alternative points of view are considered as perfectly consistent: theories including additional variables (or 'hidden variables')" (Laloë 2001, p. 656; hidden variables refers to Bohm's theory). Giere (2006a) has recognized the importance of the contingency thesis in the following terms: "… scientific knowledge claims are perspectival rather than absolutely objective. It follows almost immediately that some *contingency* is always present in any science. That human observation is perspectival, a function of an interaction between the world and human cognitive capacities seems to be indisputable" (p. 93, italics added).

Contingency at Work: Valence Bond and Molecular Orbital Models

According to Gavroglu and Simões (2012), textbooks play an important part in the early phases of the development and consolidation of subdisciplines. Pauling's (1939) *The Nature of the Chemical Bond* played a crucial role in the consolidation of the valence bond (VB) model. In this textbook/monograph, Pauling tried to convince chemists to accept quantum mechanics by illustrating that it could help to understand resonance and hybridization. Resonance was now conceptualized not as a metaphor/heuristic device/algorithm but rather as a chemical category. Similarly, resonance played an important role in understanding the hybridization of bond orbitals (Gavroglu and Simões 2012, p. 251).

For more than a decade, *The Nature of the Chemical Bond* reigned supreme until a new textbook, *Valence* (Coulson 1952), provided a new perspective of the chemical bond through the molecular orbital (MO) model. Despite Coulson's sympathies for the molecular orbital model, he considered both models to be approximations. As soon as *Valence* was published, Coulson sent a copy to Pauling, who wrote a hostile review that appeared in *Nature* (Pauling 1952). Pauling was particularly critical of Coulson's overenthusiasm for the molecular orbital method and for not having given proper credit to him for the discovery of the concept of hybridization. The *marginalia* (handwritten comments in the blank end page of Pauling's copy of *Valence*) in Pauling's copy of the book were much more critical of *Valence* and somehow did not make their way into the review (Gavroglu and Simões 2012, p. 178). Interestingly, however, Wheland (1952), Pauling's former student and long-time collaborator, not only wrote a positive review but also considered the *Valence* to be more convincing and up to date than Pauling's *The Nature of the Chemical Bond*. Given the influence of Pauling in the scientific community, in the second edition of *Valence*, Coulson (1961) incorporated most of Pauling's comments and criticisms, although he continued to have reservations with respect to resonance. Actually, he made no secret of his preference for the molecular orbital model over the valence bond model and at one stage even considered resonance to be a dirty word (Coulson 1970).

The history of the development of the covalent bond has gone through conflicting frameworks, based on the seminal work of G.N. Lewis, L. Pauling, R.S. Mulliken, and C.A. Coulson. The needs of chemists for visualization was of particular concern to Coulson who advocated *methodological pluralism* (Gavroglu and Simões 2012, p. 226), by exploring different approaches in different problems and above all emphasizing conceptual understanding instead of quantum mechanical calculations. Interestingly, Coulson reiterated that the major contribution of quantum mechanics was not that it had provided its mathematical theory but rather facilitated insight and understanding at a deeper level. Cushing (1991, pp. 337–338) in a similar vein has endorsed a framework for quantum mechanics itself, in which empirical adequacy leads to an explanation based on a set of equations and rules that finally leads to an understanding based on interpretation of the formalism.

Given this background, it is important to note that in the middle of the twentieth century, chemists had two alternative interpretations of the chemical bond (valence bond and molecular orbital models), and both were at best approximations (Niaz 2013). Furthermore, both models for presenting molecular structure were based on quantum mechanics. Pauling (1980) himself, although generally credited for having developed the valence bond model, also used the molecular orbital method for making quantum mechanical calculations. This dilemma faced by the scientific community has been expressed in cogent terms by Gavroglu and Simões (2012, p. 128):

> And, he [Pauling] claimed that what he was doing was, in effect, the theoretical justification of what Lewis, the doyen of American chemists, had already suggested so successfully nearly 20 years earlier: an explanation for the otherwise mysterious electron pair mechanism. Pauling was able to deliver. And he became the hegemonic presence of quantum chemistry, culminating in the publication of his classic *The Nature of the Chemical Bond* [...] Here one witnesses the intriguing aspects of <u>contingency at work</u>. Things, it is clear, could have developed differently. The community had different choices, both schemata [molecular orbital and valence bond] had serious empirical backing, and both schemata shared theoretical virtues (underline added).

More recently, Brush (1999) has explored the views of the chemical community based on a survey, in which authors of books or review articles on quantum chemistry or textbooks on organic chemistry or research papers on benzene or cyclobutadiene were asked the following questions: "In your opinion, which theory [MO or VB] gives the best description of benzene and similar molecules? Why?" (p. 287). Of the 133 chemists contacted, 38 replied, and the following is a distribution of their responses:

(a) For benzene, both methods are equally good ($n=10$).
(b) MO gives better results for excited states ($n=8$).
(c) MO provides a better explanation of aromaticity ($n=6$).

(d) MO better explains benzene's stability and reactions ($n = 5$).
(e) VB (modern version) gives a better description of benzene ($n = 5$).
(f) MO gives the most accurate values of heat of formation, bond lengths, etc. ($n = 2$).
(g) The ring current effect in NMR is most easily explained by MO ($n = 2$).

These results show that despite the popularity of MO, about 40 % of the respondents still consider VB to be equally good, and hence the chemical community seems to support both methods.

These considerations from the history and philosophy of science if included in general chemistry textbooks can facilitate students' conceptual understanding. Interestingly, Pauling (1980) was strongly opposed to the use of the molecular orbital model in introductory general chemistry courses and textbooks. In his opinion, valence bond model was much simpler and more powerful and also facilitated nonmathematical discussions better (for further discussion, see Garritz 2013).

Valence Bond and Molecular Orbital Theories: A Never-Ending Rivalry?

The competition between the valence bond (VB) (Pauling, Slater, and colleagues) and molecular orbital (MO) (Mulliken, Coulson, and colleagues) theories, primarily in the first half of the twentieth century, is well known in the history of science (Gavroglu and Simões 2012). Although both theories were first proposed in the 1930s, by 1955 the MO theory had become dominant. However, the rivalry between the two has continued up to almost recent days. Three practitioners of VB and MO theory have provided insight with respect to the rivalry and the underlying epistemology of chemistry (Hoffmann et al. 2003; basically a conversation between the three—a trialogue). Roald Hoffmann is a theoretical chemist at Cornell University and generally works with the MO theory. In late 1964, R.B. Woodward asked Hoffmann to make his ideas theoretically rigorous. The collaboration culminated in five papers detailing the consequences of conserving orbital symmetry in pericyclic reactions (Woodward and Hoffmann 1965) and subsequently the Woodward–Hoffmann rules. Hoffmann received the 1981 Nobel Prize in chemistry for this work. Sason Shaik works at the Lise Meitner-Minerva Center for Computational Quantum Chemistry, The Hebrew University of Jerusalem, Israel. Philippe Hiberty works at the Groupe de Chimie Théorique, Université de Paris-Sud, France. Both Shaik and Hiberty started working on VB theory in the 1980s and have provided a new impulse to the theory and hence the continued rivalry.

There are many salient features of this conversation that can help chemistry students and teachers to understand the roots of their chemical heritage. The following is a précis of the ideas expressed by the three computational chemists (RH = Roald Hoffmann, SS = Sason Shaik, PH = Philippe Hiberty; statements appear in the same order as in the original article):

(a) During the 1970s, VB theory was considered as flawed or even perhaps dead (SS).
(b) Despite some ambivalence, VB is closely tied to the chemist's bond concept (RH).
(c) VB represented the imagery and symbolism of chemical epistemology (PH).
(d) Valence bond—a portmanteau word—a stroke of genius (RH).
(e) Pauling was America's premier structural chemist; however, he ignored MO to a degree that was clearly blind and even perhaps unethical (RH).
(f) Also Pauling left chemistry and developed an interest in biology (SS).
(g) VB's lack of an explanation for the O_2 molecule was a myth. There is no evidence that Pauling may have described that O_2 was doubly bonded singlet ground state. Actually, Pauling described the molecule with two three-electron bonds and hence its paramagnetism (SS).
(h) Chemists want theories to make predictions, preferably risky ones. It was particularly for this reason that the Woodward–Hoffmann rules made such an impact (RH).
(i) MO–VB swings may have had nothing to do with "failures" of one theory and the "successes" of another. In the end, what tipped the balance in favor of MO may have been the computer implementation of MO-based theories (SS).
(j) There are people who do not believe that either orbitals (MO) or resonance (VB) structures exist and that it all depends on the densities (RH).
(k) The greater danger is that because of the facility of doing calculations these days, chemists may simply drop the entire qualitative wisdom of both VB and MO theories (SS).
(l) Seventy-odd years after the nascence of the rivalry, today's theoreticians and experimental chemists ought to know that there are two ways of describing electronic structure, which are complementary rather than exclusive of each other. If they seem different, it may be that because the truncation of the theories at their simplest form created a situation of incommensurate theories in the Kuhnian sense (SS).

This conversation between three leading practitioners of electronic structure can provide students with an overview of not only of the origins of these theories, the rivalries, and even future course of development. Students may find of particular interest that Pauling himself may have been responsible for the difficulties faced by VB in the 1940s and 1950s. Furthermore, as suggested by many general chemistry textbooks, the lack of an explanation of the O_2 molecule by VB is perhaps not correct as Pauling himself explained its paramagnetism by two three-electron bonds. A detailed discussion of the rivalry between VB and MO theories, downfall of VB, reasons for the past victory of MO, and the current resurgence of VB theory is provided by Shaik and Hiberty (2008).

Hoffmann has expressed this historical journey of the development of electronic structure in cogent terms, which provides chemistry students with insight as to how their discipline developed, based not only on the intellectual quests but also the personal vicissitudes of the scientists:

> Taken together, MO and VB theories constitute not an arsenal, but a tool kit, simple gifts from the mind to the hands of chemists. Insistence on a journey through the perfervid bounty of modern chemistry equipped with one set of tools and not the other puts one at a disadvantage. Discarding any one of the two theories undermines the intellectual heritage of chemistry. (Hoffmann et al. 2003, p. 755)

Evaluation of General Chemistry Textbooks Based on the Contingency Thesis

The objective of this study, published here for the first time, is to evaluate the degree to which general chemistry textbooks recognize the importance of contingency thesis in the development of the valence bond (VB) and molecular orbital (MO) models. The following criteria were used for the selection of textbooks:

(a) As the historical events relating to contingency thesis occurred in the 1930s, textbooks were analyzed starting from the 1960s. It was expected that about 30 years later, textbooks would include the historical details.
(b) Textbooks from different time periods, including recent ones, were selected.
(c) Based on consultations with colleagues, textbooks from the university and nearby libraries were selected.
(d) Inclusion of textbooks that have been published in various editions, which shows their acceptance by the science education community.
(e) Consultations with colleagues in different parts of the world revealed that various textbooks (especially those with various editions) selected for this study are used as translations from English. Most of these textbooks are translated into (among other languages) Spanish, Portuguese, Italian, Greek, and Turkish.
(f) Inclusion of different kinds of introductory chemistry textbooks, such as *Principles of Chemistry*; *General, Organic, and Biochemistry*; and *Chemistry for Engineering Students*. This can help teachers who write different types of textbooks that present VB and MO models at different levels.
(g) Inclusion of older editions helps to see if the textbook has changed in later editions. Also, sometimes out-of-date editions provide good examples of the subject under study.

Criteria for Evaluation of General Chemistry Textbooks

Based on previous sections, 73 general chemistry textbooks published in the USA were evaluated and classified in the following levels (see Table 6.1):

Level I Textbooks that do not mention valence bond or molecular orbital models.
Level II Textbooks that mention only the valence bond model (including VSEPR).

Table 6.1 Classification of general chemistry textbooks ($n=73$)

Classification	*n*	Textbooks
Level I	4	Goldberg (2001), Hein (1990), Stoker (1990), Raymond (2010)
Level II	18	Armstrong (2012), Bettelheim et al. (2012), Bishop (2002), Burns (1996), Daub and Seese (1996), Denniston et al. (2011), Dickson (2000), Frost et al. (2011), Hein and Arena (1997), Hill (1975), Malone (2001), McMurry et al. (2007), Quagliano and Vallarino (1969), Russo and Silver (2002), Stoker (2010), Timberlake (2010), Tro (2012), Zumdahl (1990)
Level III	37	Ander and Sonnessa (1968), Atkins and Jones (2002), Atkins and Jones (2008), Brady and Humiston (1996), Brown and Holme (2011), Brown et al. (1997), Brown et al. (2012), Chang (1998), Chang (2010), Dickerson et al. (1970), Dickerson et al. (1979), Ebbing (1996), Ebbing and Gammon (2012), Fine and Beal (1990), Hill and Petrucci (1999), Holtzclaw and Robinson (1988), Jones and Atkins (2000), Kotz et al. (2010), Lippincott et al. (1977), Mahan and Myers (1990), Masterton and Hurley (2009), Masterton and Slowinski (1977), Mcquarrie et al. (2011), Miller (1984), Moore et al. (2002), Mortimer (1983), O'Connor (1972), Oxtoby et al. (1999), Petrucci (1972), Seager and Slabaugh (2013), Segal (1989), Sisler et al. (1980), Umland and Bellama (1999), Wolfe (1988), Zumdahl (1989), Zumdahl and Zumdahl (2012), Zumdahl and Zumdahl (2014)
Level IV	14	Bodner and Pardue (1989), Brady et al. (2000), Cracolice and Peters (2012), McMurry and Fay (2001), McMurry and Fay (2012), Oxtoby et al. (1990), Oxtoby et al. (2012), Parry et al. (1970), Silberberg (2000), Spencer et al. (1999), Spencer et al. (2012), Tro (2008), Whitten et al. (1992), Whitten et al. (2013)
Level V	–	

Notes
Level I: No mention of valence bond or molecular orbital models
Level II: Mention only valence bond model (including VSEPR)
Level III: Mention both valence bond (including VSEPR) and molecular orbital models
Level IV: Provide a rationale for understanding the two models as possible explanations of molecular geometry
Level V: Consider the two models not only as approximations but also as rivals, which also implies that no single theory or model can explain all the molecular shapes observed experimentally
See Appendix 1 for list of general chemistry textbooks

Level III Textbooks that mention both the valence bond and molecular orbital models.

Level IV Textbooks that mention both the valence bond and molecular orbital models, thus explicitly providing a rationale for understanding the two models as possible explanations of the covalent bond. Both models, however, have limitations in explaining the shape of the molecules. Such presentations come quite close to understanding the two models as rivals/alternatives and if elaborated further can help to understand *contingency at work* (also pluralism). Textbooks classified in this level need to go beyond mentioning that VB and MO are alternative models.

Level V It is plausible to suggest that textbooks classified in this level would explicitly refer to the following: (a) Lewis structures help to organize a large number of chemical facts. However, these structures do not give any information about the shapes of the molecules; (b) chemists differ with respect to the relative merits of the different models (primarily VB and MO); (c) we do not have a "correct" model to predict molecular geometry; (d) no single theory or model can explain all the molecular shapes observed experimentally; (e) each model is an approximation and thus has its advantages and limitations; (f) although the MO model has been developed more extensively, it lacks the ability of VB model to facilitate greater visualization of molecular shapes; and (g) both VB and MO can be considered as rival/alternative models with different degrees of explanation.

Table 6.1 shows that only four textbooks were classified in Level I, because they do not include VB or MO models. Eighteen textbooks were classified in Level II, because they include only the VB model. None of the textbooks was classified in Level V. In what follows, examples are presented from textbooks that were classified in Levels III and IV. References to textbooks are provided in Appendix 1.

Examples of Textbooks That Were Classified in Level III

Thirty-seven textbooks were classified in Level III and the following are some examples:

> Suppose two H atoms are moving toward each other. Each atom has a single electron in a spherical 1*s* orbital. While the atoms are separated, the orbitals are independent of each other; but as the atoms get closer together, the orbitals overlap and blend to create an orbital common to both atoms called a *molecular orbital*. The two shared electrons then move throughout the overlap region but have a high probability of being found somewhere between the two nuclei. As a result, both of the positive nuclei are attracted toward the negative pair of electrons and hence toward each other. (Seager and Slabaugh 2013, p. 107)

> The VB and MO theories are both procedures for constructing approximations to the wavefunctions of electrons, but they construct these approximations in different ways. The language of valence-bond theory, in which the focus is on bonds between pairs of atoms, pervades the whole of organic chemistry, where chemists speak of σ and π bonds between particular pairs of atoms, hybridization, and resonance. However, molecular orbital theory, in which the focus is on electrons that spread throughout the nuclear framework and bind the entire collection of atoms together, has been developed far more extensively than valence-bond theory .... (Atkins and Jones 2008, pp. 116–117)

> A major weakness of the theories presented in chapters 8 and 9 [Lewis structures, VSEPR, VB] is that they do not always predict the magnetic properties of substances. An important example is O_2, which is paramagnetic. This means that O_2 molecules must have unpaired electrons. Diatomic oxygen has an even number of valence electrons (12), and the octet rule predicts that all these electrons should be paired. According to valence bond theory, O_2 should be diamagnetic, but it is not ... This discrepancy between experiment and theory for O_2 (and many others) can be *resolved by using an alternative model of covalent bonding, the molecular orbital (MO) approach. Molecular orbital theory treats bonding in terms of*

orbitals that can extend over an entire molecule. The orbitals are not confined to two atoms at a time. (Moore et al. 2002, p A.23; italics in the original)

Despite evidence to the contrary, many textbooks refer to the failure of VB theory to explain the paramagnetism of the O_2 molecule. Actually, early in the development of VB theory, Pauling (1931) had explained that this molecule consisted of two three-electron bonds that explained its paramagnetism. Such episodes in the history of chemistry provide important insight with respect to how scientists struggle to provide evidence in support of a particular theory, and this has been expressed cogently by Shaik and Hiberty (2008): "One wonders what role the animosity between the MO and VB camps played in propagating the notion of the 'failures' of VB to predict the ground state of O_2. Sadly, scientific history is determined also by human weaknesses" (p. 12). This is another good example of the role played by the social and historic milieu in the development of a theory (see Chap. 3).

Textbooks classified in Level III mention and provide examples of both valence bond and molecular orbital models. Such presentations are quite straightforward with no attempt to provide insight with respect to the underlying conflict between the VB and MO models (contingency at work) and consequently the need for further research. For example, the presentation by Seager and Slabaugh (2013) is a good preamble for introducing the MO model, but it entirely ignores that research has provided an alternative interpretation of the formation of covalent bonds. Indeed, this precisely is the point of discussion: should textbooks provide students with prescriptive accounts of "correct" scientific models or rather go a step further and introduce alternative models that are equally supported by empirical evidence? Furthermore, history of science shows that almost all models are bound to change and are fundamentally tentative in nature. Again, Moore et al. (2002) could have been classified as Level IV if they had mentioned the limitations of the MO model as a rival approach. In the next section, examples from Level IV textbooks make the difference between the two types of textbooks much more understandable.

Examples of Textbooks That Were Classified in Level IV

Fourteen textbooks were classified in Level IV and the following are some examples:

> No single theory or model yet developed succeeds in explaining all the molecular shapes observed in the laboratory. A theory that explains one group of molecules cannot explain another group. Each model has its advantages and limitations. Chemists, therefore, use them all within the areas to which they apply, fully recognizing that there is still much to learn about how atoms are assembled in molecules. (Cracolice and Peters 2012, p 380)

> Whenever two different theories are used to explain the same concept, the question comes up: Which theory is better? This question isn't easy to answer, because it depends on what is meant by "better." Valence bond theory is better because of its simplicity, but the MO theory is better because of its accuracy. Best of all, though, is a blend of the two theories that combines the strengths of both. (McMurry and Fay 2001, p 285)

> Which description is better, the delocalized molecular-orbital picture or the localized valence-bond approach? Each has its advantages and offers insights into the nature of

> chemical bonding … The truth probably lies somewhere in between, and accurate computational techniques for predicting the properties of small molecules have been developed using each model as a starting point. (Oxtoby et al. 1990, p. 765)

> Each theory [VB and MO] complements the other and is indispensable to a full understanding of covalent bonding … Don't be discouraged by our need for more than a single model to explain the observations in a topic as universal as chemical bonding. What we find in every science is that that one model accounts for a particular aspect of a topic better than another, and that several models are called into service to explain a broader range of phenomena. (Silberberg 2000, p. 396)

All textbooks classified in this level clearly point to the need for further research by recognizing that there is much to learn about how atoms combine in molecules, two different theories can be used to explain the same concept, there is no need to be discouraged if we do not have a categorical solution to a research question, and different theories combine to provide us the "truth." These textbooks provide a perspective that comes closer to that based on history of science. These are thought-provoking ideas for students starting their university career as future scientists or engineers. Indeed, it may even provide them with the incentive and the motivation for solving problems that require further research and thus contribute toward the progress of science. Examples from these textbooks with respect to VB and MO models illustrate cogently what Giere (2006a, b) has recommended: no theory can provide "a complete and literally correct picture of the world itself" (for details, see Chap. 2).

Textbooks classified in Level IV explicitly recognize the explanatory power and limitations of both the VB and MO models, which is an improvement with respect to Level III textbooks. As an example of how the two theories can be blended, McMurry and Fay provide the example of ozone (O_3). VB theory describes well the localized σ bond in ozone, and the MO theory best describes the delocalized π bond. Consequently, in the delocalized molecular orbital picture, several electrons will be found in the same region of space. On the contrary, in valence bond picture electron distribution is localized in individual bonds.

Discussion of the Structure of Benzene Based on Molecular Orbital Model

Pauling (1980) expressed his opposition to the inclusion of molecular orbital model in general chemistry textbooks in the following terms:

> It is for these reasons that I have concluded that it was a real tragedy when the writers of elementary textbooks of chemistry were so impressed by the molecular orbital method as to decide to put it into these textbooks. (p. 40)

Among other reasons, Pauling (1980) considered the molecular orbital model to be confusing even for simple molecules, such as ethylene, and much more so for benzene and concluded:

> In the latest editions of some first-year chemistry textbooks the authors do not even try to discuss benzene by the molecular orbital method, even though they have a long, not very precise discussion of molecular orbitals. (p. 40)

It is interesting to note that of the 73 textbooks analyzed in this study, 51 (70 %) present the molecular orbital model, and of these a great majority present a fairly detailed discussion of the benzene structure. The following are some of the examples of these textbooks: Umland and Bellama (1999), Holtzclaw and Robinson (1988), Chang (2010), Moore et al. (2002), Hill and Petrucci (1999), Tro (2008), Atkins and Jones (2002, 2008), Brady et al. (2000), Whitten et al. (1992), Brady and Humiston (1996), McMurry and Fay (2001), Mortimer (1983), and Mahan and Myers (1990). This finding in no way detracts from Pauling's contributions to both valence bond theory and chemistry education.

Evaluation of Organic Chemistry Textbooks

Given the importance of explaining the benzene structure in both the valence bond (VB) and molecular orbitals (MO) models, it would be interesting to study this topic in organic chemistry textbooks. Farré and Lorenzo (2012) have studied the benzene structure in five organic chemistry textbooks published in the USA, between 2000 and 2008. Without mentioning the contingency thesis, these authors have used a history and philosophy of science framework to evaluate the degree to which organic chemistry textbooks employ the VB or the MO model for explaining the benzene structure.

The benzene structure consists of six carbon atoms joined in a ring having a planar hexagonal arrangement. Each carbon atom is bonded to one hydrogen atom. Friedrich Kekulé (1829–1896) proposed such a model for the structure of benzene in 1865. At present, the structure of benzene is considered to be a hybrid of two resonance structures having alternate single and double bonds as part of a ring. The ring represents the delocalized electrons that occupy the molecular orbital. Almost all the textbooks analyzed first refer to the oscillating ring structure of benzene suggested by Kekulé in 1865, and how the classical structural theory could not satisfactorily explain its properties. Next, the authors introduce the concept of resonance as suggested by Pauling's VB approach followed by the molecular orbitals (MO) with delocalized electrons as suggested by Coulson. The authors concluded that despite the rivalry between the VB and MO approaches, both models complement each other in explaining the benzene structure. Once again, this shows the continuing rivalry between the VB and MO approaches to understanding the covalent bond (cf. Hoffmann et al. 2003).

Conclusions

1. Lewis did not make a direct contribution to models of molecular shape. However, almost all textbooks first recognize Lewis's electron-pair sharing bond and then introduce VSEPR, VB, and MO models, which shows the progressing nature of this topic in the chemistry curriculum.

2. Although none of the textbooks explicitly mention the contingency thesis, it is important to note that of the 73 textbooks analyzed, 51 did present both the VB and the MO models (most of these textbooks devote almost 50 pages to presenting these models). Consequently, it seems that the chemical community provides support for both models. The take-home message is the following: textbooks can easily present the two models within a historical perspective, especially with respect to being rival models (mention of contingency thesis itself is not essential), and this may require no more than half a page.
3. Despite Pauling's reservations, a great majority of the textbooks present the MO model. This is not to discredit Pauling but rather to recognize that both VB and MO have continued to develop different perspectives and hence the recognition of contingency.
4. Again, despite Pauling's observation of the textbooks that present the MO model, a great majority do present a fairly detailed discussion of the benzene structure based on this model.
5. Presentation of both models and a comparison of their explanatory power can provide students with a better understanding of how scientific models have to compete with their rivals (based on Hoffmann et al. 2003; Morrison and Morgan 1999; Giere 1999, 2006a, b).
6. One of the important contributions of the VB theory is to take into consideration the needs of the chemists by emphasizing visualization and understanding. Even at the early stages of the rivalry, Coulson (who favored the MO) was continually trying to find ways to appropriate quantum mechanics to the chemists' culture and at one stage considered the chemical bond to be a "concept of the imagination" (Coulson 1952b, p. 11). Interestingly, after almost 50 years, Shaik and Hiberty (2008) have tried to revive these important aspects of VB: "… this book shows that *the use of VB theory is all about insight*, and the ability of one to think, reason, and predict chemical patterns. This word insight brings to mind the Coulson admonition: 'Give me insight not numbers'" (p. 305, italics in the original). Indeed, this is what chemistry education is all about.
7. Based on a historical reconstruction of the origin of the covalent bond, valence bond (VB), and molecular orbital (MO) models, it is plausible to suggest that history and philosophy of science facilitate not only the conceptual understanding of science content but also how science progresses. In other words, this progress is perspectival rather than objective and characterized by the contingency thesis (Giere 2006a, b). Furthermore, there seems to be a relationship between Coulson's *methodological pluralism* and Giere's *perspectivism*, which shall be explored in Chap. 8 (Conclusions).
8. The transition from Lewis's cubic atom to Pauli's exclusion principle (considered to be the cornerstone of the entire science of chemistry) to the valence bond model (Pauling, Slater) to the molecular orbital model (Mulliken, Coulson) to what came next clearly shows the tentative nature of our understanding of valence.

Chapter 7
An Overview of Research in Chemistry Education

Introduction

This chapter provides an overview of research in chemistry education based on history and philosophy of science. It shows that various topics of the chemistry curriculum are the subject of research in various parts of the world. The following is a list of the topics discussed in this chapter:

1. Kinetic molecular theory of gases
   (a) Kinetic molecular theory of gases in general chemistry textbooks
2. Periodic table of the chemical elements
   (a) Periodic table in general chemistry textbooks
   (b) Teaching the periodic table in the classroom
3. Origin of the covalent bond
   (a) Origin of the covalent bond in general chemistry textbooks
4. Oil drop experiment
   (a) Oil drop experiment in general chemistry textbooks
   (b) Teaching the oil drop experiment in the laboratory/classroom
5. Electrolyte solution chemistry
   (a) Teaching electrolyte solution chemistry in the classroom
6. Photoelectric effect
   (a) Photoelectric effect in textbooks
   (b) Photoelectric effect in laboratory manuals

M. Niaz, *Chemistry Education and Contributions from History and Philosophy of Science*, Science: Philosophy, History and Education,
DOI 10.1007/978-3-319-26248-2_7

7. Wave–particle duality
   (a) Wave–particle duality in general chemistry textbooks
   (b) Wave–particle duality in the classroom

Kinetic Molecular Theory of Gases

The kinetic molecular theory of gases plays an important role in understanding many topics of the chemistry curriculum, such as atomic structure, chemical equilibrium, gases, and thermodynamics.

Kinetic Molecular Theory of Gases in General Chemistry Textbooks

Based on criteria derived from a historical reconstruction, Niaz (2000c) has analyzed general chemistry textbooks ($n=22$) published in the USA. Based on the same criteria, Niaz and Coştu (2013) have analyzed general chemistry textbooks published in Turkey. The following is a brief description of the six criteria:

Criterion 1. Maxwell's simplifying (basic) assumptions: Maxwell's assumptions (most textbooks present them), although speculative, were an attempt to reduce the complexity of the problem by introducing ceteris paribus clauses. This method helped scientists to build a series of successive theories based on a particular model of the "ideal" gas. Each tentative theory was designed to be a closer approximation to properties known to obtain in the "real" gases. This criterion is based on Achinstein (1991), Cartwright (1983), Clark (1976), Lakatos (1970), Maxwell (1860), and McMullin (1985).

Criterion 2. Inconsistent nature of Maxwell's research program: Maxwell's theory was based on the assumption that the motion of the particles was subject to Newtonian mechanics. However, at least two of the assumptions, viz., movement of the particles and the consequent generation of gas pressure, were in contradiction with Newton's hypothesis explaining the gas laws based on the repulsive forces between particles. History of science shows that many programs progressed similarly based on inconsistent foundations (cf. Bohr's program in Lakatos 1970, p. 142). This criterion is based on Achinstein (1987), Brush (1976), and Lakatos (1970).

Criterion 3. Maxwell's statistical considerations: Based on statistical considerations, Maxwell showed that the collisions of gas molecules would not simply tend to equalize all their speeds (as some had expected) but, on the contrary, would produce a range of different speeds. This consideration later led to the Maxwell–Boltzmann distribution of molecular speeds, which showed that the majority of the molecules have speeds lying within a relatively limited range, and a certain proportion of the molecules have very low and very high speeds. On increasing the temperature, the general shape of the distribution curve remains unchanged, but there is a flattening of the maximum, which now occurs at a higher speed. In other words, as the temperature increases, there is a wider distribution of speeds and the fraction of the molecules possessing high speed increases. This criterion is based on Maxwell (1860) and Porter (1981).

Criterion 4. Van der Waals' contribution: Reducing/modifying basic assumptions: If Maxwell's basic assumptions were speculative, van der Waals followed the same method by providing greater insight into Maxwell's theory. His major contribution was to reduce the assumptions in order to include the continuity of intermolecular forces, which facilitated the transition from "ideal" to "real" gases—a "progressive problem shift" (cf. Lakatos 1970). This criterion is based on Brush (1976), Clark (1976), Gavroglu (1990), and van der Waals (1873).

Criterion 5. Kinetic theory and chemical thermodynamics as rival research programs: Kinetic theory had to face from the very beginning a serious challenge from the proponents of chemical thermodynamics. This opposition was based primarily on the grounds that any theory having "arbitrary" assumptions based on invisible and undetectable atoms was beyond the fold of science. According to Lakatos (1970), history of science is a history of rival research programs. This criterion is based on Brush (1974, 1976) and Lakatos (1970).

Criterion 6. From "algorithmic mode" to "conceptual gestalt" in understanding the behavior of gases: A major contribution of Maxwell and Boltzmann was to have facilitated our understanding of gases beyond the observable, hydrodynamical laws (Boyle, Charles, Gay-Lussac) and explained the internal properties based on the kinetic molecular theory. This criterion evaluates the degree to which textbook presentation (examples, illustrations, end-of-chapter problems, etc.) explicitly recognizes that there are two modes of solving gas problems, viz., "algorithmic mode" and "conceptual gestalt." For example, in order to understand that pressure of a gas is a consequence of molecular collisions, it is not sufficient to repeat Maxwell's assumption. In order for this property of gases to be meaningful to students, it will have to be incorporated in a problem situation (cf. Item 4, Niaz and Robinson 1992). This criterion is based on Clark (1976), Hanson (1958), Coştu (2007), and Nurrenbern and Pickering (1987).

For a complete list of general chemistry textbooks published in Turkey and analyzed in this study, see Appendix 2. Table 7.1 shows that most textbooks ($n=11$) simply mentioned (M), and five textbooks described Maxwell's simplifying assumptions (Criterion 1) satisfactorily (S). Here are two examples:

> Why does a gas behave as Boyle, Charles or Gay-Lussac Laws describe? Why does a gas produce pressure? What does it mean "heat of gases"? These questions about gases were answered by the kinetic molecular theory of gases. In 1738, Bernoulli first proposed the kinetic theory of gases. After then, Clasusius, Maxwell, Boltzmann and other scientists put forward the kinetic molecular model using statistical mechanics. It is not possible to explain behaviors of gases by using derivations from directly measured properties of gases. Deducing actual characteristics of a gas from its physical properties is an approximation.

Table 7.1 Distribution of Turkish general chemistry textbooks according to criteria (kinetic molecular theory of gases) and classification ($n=22$)

Classification[a]			
Criteria	N	M	S
1	6	11	5
2	22	–	–
3	14	–	8
4	4	2	16
5	22	–	–
6	22	–	–

Note: This table is based on data presented in Niaz and Coştu (2013)
[a]*N* no mention of the criterion, *M* mention with no details, *S* satisfactory

> Critical glance on the experimental observations about behaviors of gases leads to useful approximations of a real gas [to predict approximate behaviors of a real gas]. An approximation about behavior of a gas is named as a model. Models should be containing structure and actual behaviors of them [gases]…. The model referred to as kinetic molecular theory is based on some postulates or assumptions about behavior of gases… (Baykut 1964, pp. 54–55)

> …up until now, we have examined important properties of the gases. We generalized the properties-regularity relating to gases- as gas laws. There is one question that needs to be answered. Do gases have such a structure in order to fit the properties? Scientists put forward models or theories in order to respond to such a question. The models or theories the scientists suggest are counted as viable until they truly explain related circumstances. Kinetic molecular theory is also a model (or theory) in order to explain both behaviors of gases and all facts concerning gases. Kinetic molecular theory incorporates some assumptions [or postulates] as all theories do …. (Bayın 1982, p. 93)

As seen from aforementioned explanations about the kinetic molecular theory, textbooks emphasize that kinetic molecular theory and its assumptions (or postulates) are considered to be models, approximate and tentative (models develop) in order to explain the behavior of gases. Compare this to the following example that makes a simple mention (M) of the assumptions:

> …in the previous section, we discussed how gas laws were put forward to explain empirical observations. At present, we need a theory as to why gases act in accordance with gas laws. Kinetic molecular theory by accepting that gases move in random motion successfully explains these gas laws… [after this, textbook presents Brownian motion and kinetic molecular theory] (Özcan 1998, pp. 300–301)

Of the 22 textbooks analyzed, none mentioned (N) or gave any details of the inconsistent nature of Maxwell's research program (Criterion 2, see Table 7.1). Similar to general chemistry textbooks published in the USA (Niaz 2000c), Turkish chemistry textbooks ignored that like many other programs in the history of science, Maxwell's research program although successful was also based on an inconsistent foundation.

Eight textbooks described Maxwell's statistical considerations (Criterion 3) satisfactorily (S), and the following are two examples:

In the 19th century, two famous theoretical physicists, James Clerk Maxwell and Ludwig Boltzmann, examined distributions of the molecular speeds of gas molecules. The two scientists formulated the distributions of the molecular speeds and kinetic energies of gas molecules based on statistical considerations… [they] postulated that the distributions of kinetic energies of gas molecules as shown in Fig…when the temperature of a gas is increased, average kinetic energy of the gas molecules increases. The curve broadens and shifts toward higher kinetic energies as shown in Fig…." (Bayın 1982, pp. 92–93)

One of the important and useful results derived from the kinetic molecular theory is distributions of kinetic molecular energies and molecular speeds dependent on temperature. Fig… indicates these distributions. This is called Maxwell-Boltzmann distribution. As seen from Fig…the fraction of the total number of molecules that has a particular speed is plotted against molecular speed as for three different temperatures…when the temperature of a gas is increased as in Fig…, the curve broadens and shifts toward higher speeds. Fewer molecules than previously move at the lower speeds and more molecules move at the higher speeds…" (Özcan 1998, pp. 309–310)

Table 7.1 also shows that fourteen textbooks do not mention (N) an important contribution of Maxwell and Boltzmann distribution of molecular speed. Besides, these textbooks give superficial explanations about changes in molecular speeds of gases against temperature changes without the Maxwell–Boltzmann distributions.

On Criterion 4 (Table 7.1), 16 textbooks described satisfactorily (S) van der Waals' contribution as an attempt to reduce/modify the basic assumptions. While two textbooks simply mention (M) van der Waals' contribution, four textbooks do not mention it (N). The following are two examples of satisfactory (S) descriptions:

…in 1873, Van Der Walls, a physicist, suggested that two of the postulates of the kinetic molecular theory had to be modified based on deviations of real gases from the behavior of ideal gases… Van Der Waals attributed to two reasons, as to why equation PV = nRT for ideal gases does not follow for real gases. These are: (1) The actual volume of the gas molecules, (2) Attractive forces between gas molecules… [after then, textbook gives detailed information to formulate van der Walls equation] (Baykut 1964, pp. 58–59)

The behavior of real gases deviates from the behavior of ideal gases for two reasons: (1) The kinetic theory assumes that there are no attractive forces between gas molecules, viz., it assumes that all gas molecules move freely. However, such attractions must exist in real gases and thus pressure of a real gas should be less than ideal gas. As a result, PV⟨ RT, (2) The kinetic theory also assumes that gas molecules are points in space and that the actual volume of the gas molecules is not significant. However, this does not hold for real gases and thus PV⟩ RT. Because of the two derivations, Van Der Waals, a Dutch physicist, corrected the equation of state for an ideal gas based on the two effects… (Şenvar 1989, p. 54)

Some textbooks, although classified as Satisfactory (S), simply mentioned that van der Waals modified/corrected the ideal gas equation, without any reference to the tentativeness of the simplifying assumptions. This was done on the ground that these textbooks gave a fairly detailed, step-by-step description, of the two corrections by van der Waals. Nevertheless, it is important to emphasize that most textbooks that were classified as Satisfactory (S) do not conceptualize van der Waals' contribution as an attempt to modify Maxwell's simplifying assumptions (ceteris paribus clauses), which led to a "progressive problem shift" (Lakatos 1970).

None of the textbooks described satisfactorily (S) or briefly mentioned (M) the historical background that led to the rivalry between the research programs of the kinetic theory and chemical thermodynamics (Criterion 5).

On Criterion 6, none of the textbooks described satisfactorily (S) or briefly mentioned (M) the two modes of solving gas problems, viz., the algorithmic mode and that of conceptual understanding. Almost all of the textbooks focused mainly on problem-solving as an algorithmic mode. These textbooks generally present problems that require mathematical calculations. Two typical examples of such problems are presented here:

> Calculate the u_{rms} speed, in m/s, for H_2 at 50 °C? (Soydan and Saraç 1998, p. 155)

> Calculate the pressure exerted by 142.0 g of $Cl_{2\,(g)}$ confined to a volume of 5.0 liter at 25 °C, using both ideal gas laws and the van der Waals equation? (Alpaydın and Şimşek 2006, p. 162)

It is concluded that very few Turkish general chemistry textbooks presented the development of the kinetic molecular theory within a historical perspective. Similar results have been reported for general chemistry textbooks published in the USA (Niaz 2000c). The study (Niaz and Coştu 2013) reported here can help to improve textbooks published not only in Turkey but also in other countries. It can provide students with a historical perspective based on the development of scientific theories involving controversies, conflicts, and rivalries among scientists that is science as a human enterprise. Moreover, it may encourage some chemistry textbook authors to become interested in research on history and philosophy of science and facilitate teachers' conceptual understanding of the kinetic molecular theory of gases.

Periodic Table of the Chemical Elements

The periodic table forms an important part of almost all chemistry curricula and textbooks published worldwide. Most chemistry teachers consider the periodic table to be an important concept, both in principle and practice. Despite its importance and long history, the periodic table is still considered by students to be a difficult topic. Interestingly, Mendeleev's textbook (*Principles of Chemistry*, written between 1868 and 1870) was an endeavor to facilitate students' understanding of methods of observation, experimental facts, laws of chemistry, and, perhaps the most important of all, the "… unchangeable substratum underlying the various forms of matter" (Mendeleev 1897, Preface, p. vii). Indeed, this "unchangeable substratum" represents Mendeleev's fundamental presupposition with respect to the periodicity of properties in the periodic table as a function of the atomic theory. In spite of the long history of the periodic table and its relevance for chemistry and chemistry education, historians and philosophers of science are still trying to understand its origin, nature, and development (Bensaude-Vincent 1986; Brush 1996;

Gordin 2004; Weisberg 2007). On the other hand, although the periodic table is an important topic of research in chemistry education (e.g., Demircioğlu et al. 2009), little of this research has been conducted within a history and philosophy of science framework.

Periodic Table in General Chemistry Textbooks

Based on a history and philosophy of science framework, Brtio et al. (2005) have analyzed 57 general chemistry textbooks published (between 1966 and 2002) in the USA. The following criteria were used to analyze these textbooks:

1. *The importance of accommodation in the development of the periodic table.* Accommodation (in periods and groups) of the different chemical elements in the periodic table, according to their physical and chemical properties, is considered an important factor in the success and acceptance of the periodic table (Brush 1996; van Spronsen 1969).
2. *The importance of prediction as evidence to support the periodic law.* After the discovery of gallium in 1875, chemists devoted more attention to the periodic law, and the table was increasingly recognized as an important tool for both education and research (Brush 1996; van Spronsern 1969; Weisberg 2007).
3. *Relative importance of accommodation and prediction in the development of the periodic table.* There is considerable controversy among historians and philosophers of science with respect to the relative importance of accommodation and prediction (Brush 1996; Lipton 2005a, b; Maher 1988). For chemistry education, it is important to note that the success of the periodic table could be attributed to accommodations, predictions, or both.
4. *The role of novel predictions.* Brush (1996) considers the correction of various atomic weights by Mendeleev as novel predictions. For example, in the case of Be, he accepted 9 instead of 14, U 240 instead of 120, and Te 125 instead of 128.
5. *Explanation of periodicity in the periodic table.* How does one explain the periodicity of the elements in the development of the periodic table? The idea behind this criterion is to make students aware that, before the electronic structure of the atom was discovered, different explanations were offered for periodicity. A historical reconstruction shows that this is a controversial issue and generally two alternatives are presented: (a) inductive generalization and (b) periodicity as a function of the atomic theory, that is, before the electronic configurations were definitely elaborated (Bohr 1913; Brush 1996; Lewis 1923; Mendeleev 1879, 1889; Moseley 1913, 1914; Niaz et al. 2004: Thomson 1897; van Spronsen 1969).
6. *Mendeleev's contribution: Theory or an empirical law?* This criterion tries to analyze the nature of Mendeleev's contribution and, hence, facilitates understanding of scientific progress. Given the controversy among philosophers of science, a historical reconstruction provides three alternatives: (a) an ordered domain or codification scheme, (b) an empirical law, and (c) a theory with limited

explanatory power or an interpretative theory. This reconstruction is based on Cartwright (1983), Giere (1999, 2006a), Lakatos (1970), Shapere (1977), Wartofsky (1968), Weisberg (2007), and Ziman (1978). The theoretical status of the periodic table continues to be a subject of interest in recent history and philosophy of science literature (cf. Drago 2014; Gordin 2012; Niaz et al. 2004).

7. *Development of the periodic table as a progressive sequence of heuristic principles.* A historical reconstruction of the periodic table shows that it can be understood as progressive sequence of heuristic principles based on the following contributions: (a) early ideas about atomic theory and accumulation of data with respect to the atomic weights of the elements and their properties; (b) first attempt to classify elements by Döbereiner and later by De Chancourtois, Odling, Meyer, Newlands, Hinrichs, and other attempts before Mendeleev; (c) Mendeleev's first periodic table in 1869 based on atomic weights and subsequent contributions; (d) discovery of argon in 1895 and its accommodation in the periodic table; and (e) contribution of Moseley (1913) and the modern periodic table based on atomic numbers (Brush 1996; Lakatos 1970; van Spronsen 1969).

Based on these criteria, textbooks were classified as Satisfactory (S), if the textbook explicitly refers to the underlying issues in the criterion; Mention (M), a simple mention of the issues involved; and No mention (N), no reference is made to the issues involved. Results obtained showed that on Criterion 1, of the 57 textbooks analyzed, 55 presented a Satisfactory (S) description of the importance of accommodation of the elements according to their physical and chemical properties. Most textbooks devoted 50 or more pages, including color photographs and even three-dimensional figures. On Criterion 2, 30 textbooks emphasized the importance of prediction satisfactorily (S) as evidence to support the periodic law. On Criterion 3, none of the textbooks was classified as Satisfactory (S). This shows that textbooks do recognize the importance of accommodations (Criterion 1) and predictions (Criterion 2), but not the relative importance of the two (Criterion 3). It is of interest to note that, in contrast to Criterion 2 (prediction), only 10 textbooks were classified as Satisfactory (S) on Criterion 4 (novel prediction). Apparently, textbooks gave more importance to predictions (unknown elements) than novel predictions (known elements whose atomic weights were corrected). On Criterion 5, none of the textbooks was classified as Satisfactory (S). Of the 14 textbooks that were classified as Mention (M), the following are two examples:

> Mendeleev's approach to the periodic table was *empirical*; he based his classification scheme on the observed facts. (Hill and Petrucci 1999, p. 316, italics in the original)

> Early in the nineteenth century, when Dalton's atomic theory was winning general acceptance, the first attempts were made toward classification of the elements into groups or families on the basis of similarities of physical and chemical properties … even in its primitive form as stated in 1869, this [periodic] law clearly pointed to regularities that hinted at an orderly subatomic structure of matter and provided a tremendous stimulus toward seeking to understand the internal structure of atoms, as chemists and physicists sought to construct an atomic model that would explain Mendeleev's generalization (Sisler et al. 1980, p. 150)

Although both of these textbooks were classified as Mention (as they do refer to different approaches to periodicity), the difference between the two is striking. The presentation by Hill and Petrucci (1999) is laconic and at best quite simplistic. On the other hand, the presentation by Sisler et al. (1980) makes an attempt to engage students with the historical context and thus provide a deeper understanding of how the "internal structure of atoms" provided the explanation for periodicity. Furthermore, this is a good illustration of how the history of chemistry is already "inside" chemistry (Niaz and Rodríguez 2001), provided we make an effort to facilitate students' conceptual understanding beyond that of the simple regurgitation of experimental or empirical details.

Once again, on Criterion 6 none of the textbooks was classified as Satisfactory (S) and 5 were classified as Mention (M), and the following is an example:

> Indeed, the periodic table is considered to be the single most useful study aid available for organizing information about the elements. *For many years after the formulation of the periodic law and the periodic table, both were considered to be empirical.* The law worked and the table was very useful, but there was no explanation available for the law or for why the periodic table had the shape it had. It is now known that the theoretical basis for both the periodic law and the periodic table lies in electronic theory. (Stoker 1990, p. 155, italics added)

A major problem with this presentation is that it gives students an impression that for almost 100 years (approx. 1820, Dalton to 1920, Moseley) nobody asked or wondered as to why the periodic table worked. On the contrary, the historical record (including Mendeleev, Meyer, and others) shows that many scientists were constantly struggling to provide an explanation. Furthermore, it ignores that science is progressive and scientists are continuously trying to provide better explanations, namely, the tentative nature of science.

The periodic table provides a good opportunity to "weave" the various heuristic principles to provide a semblance of a sequence in the form of a convincing argument (the subject of Criterion 7). None of the textbooks was classified as Satisfactory (S) on Criterion 7, and 30 were classified as Mention (M). For example, one textbook (Phillips et al. 2000) takes almost 1.5 pages to explain Döbereiner's triads (early nineteenth century) and showed that the properties of the elements had a relationship to their atomic mass. This clearly shows that in some cases textbooks go to considerable length to illustrate some historical episodes and at the same time ignore many others and thus do not provide a balanced historical reconstruction.

At this stage it is interesting to consider the presentation of the periodic table in chemistry textbooks published in a different culture and language. Mehlecke et al. (2012) have analyzed five secondary school chemistry textbooks written in Portuguese and published in Brazil. These authors used exactly the same seven criteria based on history and philosophy of science (HPS) as developed by Brito et al. (2005). On each criterion, all textbooks were classified as N=No mention, M=Mention, and S=Satisfactory. Furthermore, these textbooks were sponsored by the National Program for Intermediate Education Didactic Book (PNLEM) and distributed in all Brazilian public schools (El-Hani et al. 2005). Results obtained are presented in Table 7.2.

Table 7.2 Distribution of Brazilian chemistry textbooks according to criteria (periodic table) and classifications ($n=5$)

Classification[a]			
Criteria	N	M	S
1	0	1	4
2	0	1	4
3	5	0	0
4	3	2	0
5	4	1	0
6	0	5	0
7	4	1	0

[a]*N* no mention of the criterion, *M* mention with no details, *S* satisfactory

Similar to the textbooks published in the USA (Brito et al. 2005), most Brazilian textbooks did fairly well on criteria 1 and 2, which dealt with empirical aspects (accommodation and prediction) of the periodic table. None of the Brazilian textbook was classified as Satisfactory (S) on criteria 3, 4, 5, 6, and 7, which is again quite similar to the textbooks published in the USA. These criteria precisely needed reflection, reasoning, and conceptual understanding. Now, let us consider the presentation from a Brazilian textbook that was classified as Mention (M) on Criterion 5 (explanation of periodicity):

> Despite all the precision, this work [periodic table] was just based on empirical knowledge based on properties of substances that was available in that epoch. In Mendeleev's epoch it was not possible to explain the cause of periodicity in the physical and chemical properties of the elements. The first atom models—Thomson's model and Rutherford's model—also faced this lacuna. Later the existence of discrete energy levels for the electrons and that the atoms of two elements in the same period of the periodic table have their most energetic electrons occupying the same energy level—Bohr's model made it possible to explain the periodicity of various atomic properties associated with the physical and chemical properties of substances based on the distribution of electrons in different levels. (Mortimer and Machado 2005, p. 111; reproduced in Mehlecke et al. 2012, p. 538)

Now, let us compare this presentation with that of the following textbook published in the USA that was also classified as Mention (M):

> In the 1800s, the atomic theory captured the imagination of chemists … many new elements were discovered and added to the list of previously known elements. Information about elements began to accumulate. During these times, scientists became aware that the properties of some elements were very similar. Eventually, they noticed that some periodic or repeating pattern of properties existed among the elements … Observations of the similarities and differences in the behavior of elements stimulated the curiosity of many chemists. Was there a grand pattern to such similarities? If such a pattern existed, what message did it convey about the nature of matter? … Today, however, it is recognized that the periodic behavior is related to increasing atomic numbers rather than atomic weights. Mendeleev's table represented a revolutionary step in the development of chemical science. (Dickson 2000, pp. 121–122)

Firstly, despite some differences, both presentations (Mortimer and Machado 2005, Brazilian textbook; Dickson 2000, USA textbook) have several common features: (a) Mendeleev's and earlier periodic tables were essentially based on empirical knowledge. (b) Scientists found that many chemical elements had several similar properties. (c) Was there an underlying tendency to explain these periodic properties? (d) In Mendeleev's time it was not possible to explain the cause of periodicity. (e) It was the modern atomic structure, based on the work of Bohr and others, that helped to explain the cause of periodicity, by replacing atomic weight with atomic number. At this stage, a student may be perplexed to note that for almost 100 years (1820–1920) scientists made no effort to understand the underlying pattern of periodicity. Secondly, the similarities between the two books, published in different languages and cultures, do provide an underlying pattern for understanding progress in science that is essentially empiricist or what Holton (1969) has referred to as the "experimenticist fallacy." Thirdly, the role played by the atomic theory in the development of first the atomic weights and other related concepts such as valence is almost completely neglected.

In light of the historical reconstruction presented in this section, to state that the periodic table was based entirely on empirical knowledge and that Mendeleev had no theory or model to explain the periodicity of the properties of the elements is perhaps rather simplistic and difficult to sustain. It is more fruitful and plausible to present a more balanced picture to the students by highlighting the dilemma (going beyond the observables) faced by Mendeleev and others in which they endeavored to look for underlying patterns to explain and understand periodicity. Here is an example from a textbook that can provide guidelines with respect to issues discussed in this section, for future students and textbook authors:

> In many ways, the creation of the periodic table by Dmitri Mendeleev in 1869 is an ideal example of how a scientific theory comes into being. At first, there is only random information—a large number of elements and many observations about their properties and behavior. As more and more facts become known, people try to organize the data in ways that make sense, until *ultimately a consistent hypothesis emerges*. (McMurry and Fay 2001, p. 160, italics added)

From a HPS perspective, one may not agree with some aspects of this presentation. Nevertheless, the recognition of the role played by "emerging hypotheses" can facilitate a better understanding of the challenges faced by Mendeleev and others, in their struggle to go beyond the observable entities. With this background the stage is set for introducing HPS while teaching the periodic table in the classroom, which is the subject of the next section.

Teaching About the Periodic Table in the Classroom

The presentation of the periodic table in most general chemistry textbooks does not facilitate a conceptual understanding with respect to its origin and development. Furthermore, these textbooks ignore that Mendeleev not only presented the periodic law to construct the periodic table but also:

(a) "Speculated" with respect to the possible cause of periodicity
(b) Hypothesized with respect to the structure of the atom long before J.J. Thomson started his experiments on cathode rays in 1897

Moore (2003) has strongly endorsed the use of the history of the periodic table in the classroom:

> Asking students to argue pro or con for a particular representation of periodicity can be a challenging and instructive exercise. It requires that they know enough about properties of the elements to make convincing arguments, and it points out that science does not always arrive at a single, best, and correct answer to a complicated question. (p. 847)

Interestingly, the presentation of this topic in most textbooks and curricula makes students think that science unequivocally provides "single," "best," and "correct answers." Based on these considerations, Niaz and Luiggi (2014) have designed a teaching strategy based on HPS aspects in order to facilitate undergraduate students' conceptual understanding of the periodic table.

This study is based on two groups of undergraduate students enrolled in a Chemistry I course at a major university in Latin America. Students in the *control group* ($n=45$) were asked to first look for information about the periodic table on the Internet and in the following textbooks: Chang (2007) and Mahan and Myers (1990). Next, the instructor based on a traditional expository method presented the following aspects: early periodic systems (Döbereiner, Newlands, and Meyer, among others), Mendeleev's periodic table, classification and order of the elements, corrections in atomic weights, predictions of new elements, and contribution of Moseley (from atomic weights to atomic numbers). This phase of the study lasted about two weeks. A week later students were evaluated on Posttest 1 and later evaluated after about 2–3 weeks on Posttest 2.

Students in the *experimental group* ($n=32$) were also asked to look for information about the periodic table in the Internet and the following textbooks: Chang (2007) and Mahan and Myers (1990). Next the instructor presented and discussed the same aspects of the periodic table that were used with the control group. A novel feature of this presentation was that it included considerations from the history and philosophy of science, such as (based on Brito et al. 2005): (a) importance of accommodation of chemical elements according to their properties in the periodic table; (b) importance of prediction of new elements as evidence for the periodic law; (c) relative importance of accommodation and prediction in the development of the periodic table; (d) illustrations of periodicity in the periodic table; (e) contribution of Mendeleev: theory or an empirical law? (f) development of the periodic table as a progressive sequence of heuristic principles: early ideas about atomic theory (e.g., Dalton) → attempts to classify elements starting in 1817 → Mendeleev's first periodic table in 1869 → discovery of argon in 1895 → contribution of Moseley in 1913 based on atomic numbers.

After this presentation, students were asked to construct concept maps (Novak, 1990). Students had experience in the construction of concept maps as part of a course "Cognitive Development and Learning Strategies," in the previous semester. In the following week, students were evaluated through Posttest 1. Based on students' responses to the four items of Posttest 1, classroom discussions helped to clarify different aspects of the periodic table. Next, in order to motivate students, a PowerPoint presentation of various historical episodes was presented by the instructor.

After this, students were invited to construct concept maps again, which were discussed in class and compared to their previous concept maps. All these activities lasted about 3–4 weeks, at the end of which students were evaluated on Posttest 2. Finally, five volunteer students participated in semi-structured interviews, each of which lasted about 45 min. Both groups of students were tested on the following Posttests:

Posttest 1

Item 1: In your opinion, what was the criterion used by Mendeleev to put the elements in the established order in the periodic table?
Item 2: If the periodic table was elaborated before the modern atomic theory, do you think there is a relationship between the periodic table and the earlier atomic theory?
Item 3: If the periodic table was elaborated before the modern atomic theory, how could Mendeleev and others construct the periodic table?
Item 4: Did the idea of ordering the elements originate with Mendeleev's periodic table?

Posttest 2

Item 5: In your opinion, in the acceptance of the periodic table, which of the following factors was most important?

(a) Accommodation of the chemical elements that is classification according to their physicochemical properties
(b) Prediction of some of the elements that were discovered later
(c) Corrections of the atomic weights of some of the elements
(d) No/ambiguous response

Item 6: In your opinion, which factors were important for the development of the periodic table? (Note: In this item students generated their own factors, which are presented in the Results and Discussion section).
Item 7: Periodicity of elements in the periodic table is: A consequence of physically observable properties (as aggregates) or chemical atoms as particles?

Students' responses were classified as conceptual or rhetorical. A conceptual response showed an understanding of the underlying issues, whereas a rhetorical response simply reiterated the information provided. Results obtained revealed that the students in the experimental group provided conceptual responses on all items. Item 1 dealt with atomic theory as the criterion used by Mendeleev to order the elements, and 19 % of the students responded conceptually. Item 2 dealt with the relationship between the periodic table and the early atomic theory, and 47 % of the students responded conceptually. Item 3 dealt with the question as to how Mendeleev could elaborate the periodic table before the modern atomic theory, and 28 % of the students responded conceptually. Item 4 asked if the idea of ordering the elements originated with Mendeleev and 13 % of the students responded conceptually. Item 7 referred to periodicity as a function of the chemical atoms (atomic theory) and 13 % of the students responded conceptually. Apparently, Items 1 and 7 refer to the same conceptual aspect, and still the percentage of students responding conceptually decreased from 19 % to 13 %. It seems that "periodicity as a function of chemical atoms" in Item 7 was more difficult to understand than the "criterion used by Mendeleev to order the elements" in Item 1. Here is an example of a conceptual response from the experimental group on Item 7:

> In the beginning the periodicity of the elements was studied by Mendeleev according to the atomic weight and physicochemical properties. Later these classifications were corrected by the valence and electron configurations [Bohr, Moseley] of the elements. At this stage it is important to clarify that the *physicochemical properties are a function of the atomic or particulate nature of the elements*, which is in turn manifested by valence and electron configuration. (Reproduced in Niaz and Luiggi 2014, p. 30, italics added)

Items 5 and 6 were slightly different, as the responses to these were not classified as conceptual or rhetorical. Nevertheless, a comparison of performance on Items 5 and 6 provides interesting insight into students' thinking and understanding. On Item 5, 34 % of students in the experimental group selected option (a), that is, accommodation of the elements, and 28 % selected option (c), that is, corrections of atomic weights. In contrast, on Item 6, where the students could provide their own factors, entirely new options appeared, such as b) Dalton's atomic theory (16 %); (c) Karlsruhe Congress (13 %); and (d) placement of the noble gases (6 %), which gives a total of 35 %. This clearly shows that given the opportunity, students in the experimental group can go beyond the factors discussed in the traditional classrooms and textbooks.

As students in the control group were not exposed to the treatment, it was not expected that they would respond conceptually. Nevertheless, one student responded conceptually to Item 1, two students responded conceptually to Item 2, and one student responded conceptually to Item 3. How can we explain conceptual responses by students in the control group who received instruction in a previous semester and hence could not have interacted with the students in the experimental group? In order to respond to this question, let us analyze the response provided by student #33 of the control group to Item 2, which has the following critical aspects: (a) A clear distinction between the early atomic theory and the modern atomic theory. (b) Some properties of the atoms were known quite early. (c) Early chemists must have had some notions of the atomic theory based on properties of the atoms. (d) Study of the atoms led to the study of the physical and chemical properties of the elements and subsequently their ordering in the periodic table. All these four critical aspects are in general discussed in almost all general chemistry textbooks, namely, Dalton's atomic theory; physical and chemical properties of the elements and their compounds; contributions of Gay-Lussac, Avogadro, and others; and of course the early attempts to order the elements as early as 1817 by Döbereiner. It is plausible to suggest that given the opportunity to reflect and the appropriate test format (as provided by this study, through Items 1–7), at least some students could establish a relationship between the early atomic theory, properties of the elements, and their ordering in a periodic table. Conceptual responses by control group students provide a good argument for including such material in classroom discussions, especially in the context of the periodic table.

In conclusion, it is suggested that while teaching the periodic table both at the high school and introductory university courses, the following can constitute guiding principles:

1. How could a simple arrangement of the elements based on atomic mass (atomic weight for Mendeleev) provide such regularities as observed in the periodic table?
2. Many scientists including Mendeleev were continually trying to understand the underlying reason for periodicity, and various attempts were made to understand and classify the elements. On the contrary, most textbooks give the impression that for almost 100 years (1820–1920), scientists had no idea or never asked the question as to whether there could be an underlying rationale for explaining periodicity.
3. Besides Mendeleev in 1869, the following codiscoverers of the periodic table also made important contributions: De Chancourtois in 1862, Odling in 1864, Meyer in 1864, Newlands in 1865, and Hinrichs in 1866.
4. Even before the modern atomic theory (proposed in 1897), scientists were well aware that periodicity in the periodic table is a function of the atomic theory.
5. Accommodations and predictions of elements provided important evidence for the acceptance of the periodic law, and it would be helpful to emphasize both in the classroom.
6. Based on a historical reconstruction, the following sequence of heuristic principles can help to facilitate understanding: accumulation of atomic weights of the elements in the early nineteenth century, attempts to classify elements starting in 1817, Karlsruhe congress in 1860, Cannizaro's contributions, Mendeleev's first periodic table in 1869, corrections of known atomic weights, discovery of argon in 1895, and contribution of Moseley in 1913 that led to the periodic table being based on atomic numbers.
7. Implementation of these guiding principles constitutes what in the history of science and science education literature has been referred to as "science in the making" (for details, see Niaz 2012a).
8. An effective way in which to bridge the gap between how we teach science (periodic table in this case) and what scientists actually do, that is, "science in the making," is through the inclusion of humanizing aspects of the history of science in the form of a story (contextual teaching; cf. Klassen 2006).

Origin of the Covalent Bond

Chemical bonding and the introduction of covalent bonds is considered to be a difficult topic for most high school and freshman students (Gillespie et al. 1996; Ünal et al. 2006, 2010). Ionic bonds are formed by the actual transfer of electrons, which produces positively and negatively charged ions. Formation of the ionic bond leads to the lowering of energy because of electrostatic attraction between ions of opposite charge. For example, in the case of sodium chloride, "An ionic solid consists of an array of enormous numbers of cations and anions stacked together to give the lowest energy arrangement" (Atkins and Jones 2008, p. 58). In this context, how can we explain the lowering of energy when two electrons are shared to form a covalent bond? Apparently, the approach of two electrons having the same charge should produce repulsive forces and hence result in destabilization. Most students if given an opportunity to think and reflect can be perplexed by this dilemma. It is not surprising that when first proposed the idea of a covalent bond was considered by some leading scientists to be not only untenable but even "absurd" and "bizarre."

Origin of the Covalent Bond in General Chemistry Textbooks

Niaz (2001c) has presented a historical reconstruction of the events that led to the postulation of the covalent bond by G.N. Lewis. Based on criteria derived from this reconstruction, Niaz (2001c) evaluated general chemistry textbooks (published in the USA) and found that most textbooks lacked a history and philosophy of science (HPS) perspective and thus did not deal adequately with the dilemma faced by the students. Furthermore, in an attempt to simplify the topic, most textbooks present rules (algorithms, 5–10 pages) for writing Lewis diagrams for covalent bonds, which are memorized by the students and do not facilitate conceptual understanding.

The following criteria were used for analyzing general chemistry textbooks published in Turkey (Niaz and Coştu 2013), based on a historical reconstruction of the origin of the covalent bond presented earlier by Niaz (2001c):

Criterion 1. Lewis's cubic atom as a theoretical device for understanding the sharing of electrons: Lewis's cubic atom was based on his atomic theory based on postulates formulated in 1902. The cubic atom was thus a theoretical device that was later used for understanding the sharing of electrons (covalent bond) and provided the rationale for the octet rule. This criterion is based on the following references: Jensen (1984), Kohler (1971), and Lewis (1916, 1923). The following classifications were used: *Satisfactory (S)* (treatment of the subject in the textbook is considered to be satisfactory if it is briefly explained that Lewis (1916) used his model of the cubic atom to explain the sharing of electrons and the octet rule), *Mention (M)* (a simple mention of Lewis's cubic atom), and *No mention (N)* (no mention of Lewis's cubic atom).

Criterion 2. Sharing of electrons (covalent bond) had to compete with the transfer of electrons (ionic bond): Lewis's idea of sharing electrons (covalent bond) had to compete with the transfer of electrons (ionic bond). The origin of the ionic bond as the dominant paradigm in chemical combination can be traced back to Thomson's discovery of the electron in 1897 (according to Arabatzis (2006), it is misleading to attribute the discovery of the electron exclusively to J.J. Thomson). By 1913, the ionic bond theory completely dominated chemistry, and it was in the early 1920s that Lewis's idea of sharing electrons became acceptable. This criterion is based on the following references: Kohler (1971), Lakatos (1970), Lewis (1916, 1923), and Thomson (1897, 1907, 1914) The following classifications were used: *Satisfactory (S)* (treatment of the subject is considered to be satisfactory if the role of competing frameworks (polar/nonpolar) is briefly described), *Mention (M)* (a simple mention of the competing frameworks), and *No mention (N)* (no mention of the competing frameworks).

Criterion 3. Covalent bond: inductive generalization/derived from the cubical atom: The objective of this criterion (Kohler 1971; Lakatos 1970; Rodebush 1928) is to evaluate if the textbooks follow one of the following interpretations with respect to the origin of the covalent bond (shared pair): *Inductivist (I):* Lewis's covalent bond was based on: stability of the noble gases or formation of the hydrogen molecule leads to a lowering of the energy, or helium an inert gas has a pair of electrons, or numbers of electrons in most compounds are even. *Lakatosian (L)*: Lewis's (shared pair) covalent bond was not induced from experimental evidence but derived from the cubic atom. *No mention (N)*: textbook makes no mention explicitly to either of the two interpretations, presented above.

Criterion 4. Pauli's exclusion principle as an explanation of the sharing of electrons in covalent bonds: The objective of this criterion is to evaluate if textbooks consider Pauli's exclusion principle (before the concept of spin) to provide an explanation of the sharing of electrons. This criterion is based on the following references: Kohler (1971), Lakatos (1970), Pauli (1925), and Rodebush (1928). The following classifications were elaborated: *Satisfactory (S)* (treatment of the subject in the textbook is considered to be satisfactory if the role of Pauli's exclusion principle is briefly described, in order to explain the covalent bond), *Mention (M)* (a simple mention of Pauli's exclusion principle, in the context of the covalent bond), and *No mention (N)* (no mention of Pauli's exclusion principle).

For a complete list of general chemistry textbooks published in Turkey, analyzed in this study, see Appendix 2. On Criterion 1 (Table 7.3) none of the textbooks described satisfactorily (S) or mentioned (M) Lewis's cubic atom within a history

Table 7.3 Distribution of Turkish general chemistry textbooks according to criteria (origin of the covalent bond) and classification ($n=27$), based on Niaz and Coştu (2013)

Classification[a]			
Criteria	N	M	S
1	27	–	–
2	24	3	–
4	22	3	2
	I	L	N
3	19	–	8

[a]*N* no mention of the criterion, *M* mention with no details, *S* satisfactory, *I* inductivist, *L* Lakatosian

and philosophy of science (HPS) framework. Similarly, in a recent study Croft and De Berg (2014) have reported that, of the eight secondary school chemistry textbooks analyzed, none referred to Lewis's cubic atom.

None of the textbooks described satisfactorily (S) that Lewis's idea of sharing of electrons (covalent bond) had to compete with the transfer of electrons, that is, the ionic bond (Criterion 2). Only three textbooks made a simple mention (M) and the following are two examples:

> Previously, it was commonly accepted that all chemical bonds can form between ions through electrostatic attractions, that is, it was accepted that all chemical bonds were ionic bonds. However, in 1906 [1916], American chemist G. N. Lewis said that in some cases, the idea that electrons transfer entirely from one atom to another atom was illogical… [Comment: Textbook provides the example of formation of H_2 to rebut ionic bond theory. However, it does not explicitly interpret the origin of the covalent bond as a rival research program based on an HPS perspective.] (Aydın et al. 2001, pp. 73–74). (Note: Lewis published his ideas in 1916)

> Examining the ionic bond, we saw a bond formed by transfer of one or more electrons between two atoms, whose electron affinity and ionization energies were very different. In a wide variety of cases, a more stable state did not form with ionic bonding. On the contrary, a more stable state formed with covalent bonding between two atoms whose electron affinity and ionization energies were identical. As an example, consider the bond formed by two hydrogen atoms… [textbook explains formation of the hydrogen molecule in detail]… in the formation of this bond [H-H], electron transfer from one atom to the other is impossible [textbook gives detailed reasons, implying rebuttal of the ionic bond] … therefore, covalent bond is formed differently as compared to ionic bonding… [Comment: Textbook also provides detailed information in the following paragraphs, implying rebuttal of ionic bonding. However, the textbook does not explicitly interpret the origin of the covalent bond as a rival research program based on an HPS perspective.] (Özcan 1998, pp. 184–185)

Most textbooks ($n = 24$) made no mention (N) that Lewis's idea of sharing of electrons (covalent bond) had to compete with the transfer of electrons (ionic bond). The controversial origin of the covalent bond and its rivalry with the ionic bond provides a good opportunity to illustrate how progress in science is based on controversy and how established theories or ways of thinking are difficult to change. Here is an example of a textbook that was classified as (N) and shows the difference between textbooks classified as (M):

> Covalent bonds are bonds between two identical or different non-metals. Since the electronegativity of two atoms is close to each other, there is little difference in the abilities of two atoms to attract the bonding electrons to them. Therefore, electron transfer between two atoms does not occur; instead, the electrons involved in such a bond are shared. A chemical bond formed by sharing electrons is called covalent bond… (Alpaydın and Şimşek 2006, p. 93)

As seen from the evaluation of textbooks, they do not interpret the origin of the covalent bond as a rival research program, based on an HPS framework (Lakatos 1970). Besides, the textbooks only provide students with detailed information for writing the Lewis structures. Even a brief mention of the historical details can facilitate conceptual understanding of the difference between ionic and covalent bonds.

On Criterion 3 (Table 7.3), none of the textbooks presented the Lakatosian interpretation (L), viz., tracing the origin of the stability of the covalent bond to the cubic atom and giving enough details to show that Lewis's ideas developed slowly based on conjectures. Most textbooks ($n=19$) consider the origin of the covalent bond to be an inductive (I) generalization and the following is one example:

> … between two identical atoms, ionic bonds cannot be formed. Therefore, how is a bond formed between such atoms? The question was answered in 1916 by the American chemist Gilbert N. Lewis…G.N. Lewis supposed that the bond [between two identical atoms] is a covalent bond…. [textbook gives additional information about covalent bonds and an example of H_2 molecule]…as a result of filled in outer shell of the atom with shared electron, a bond between two atoms leads to stable molecules if they share electrons in such a way as to create a noble gas configuration for each atom as shown Figure… [one page later, textbook gives following explanation dealing with inductive generalization] …Helium does not form a molecule of He_2, because repulsive forces exert on attractive forces as distance between the two helium atoms decreases. Therefore, the atoms do not come near enough to form a bond… (Bayin 1982, p. 226)

This presentation is quite representative of most textbooks and shows explicitly that the octet rule is sustained by empirical evidence. On the other hand, eight textbooks made no mention explicitly to either of the two interpretations (Lakatosian or inductive generalization).

Table 7.3 (Criterion 4) shows that three textbooks mentioned (M) and only two textbooks described satisfactorily (S) Pauli's exclusion principle as an explanation of the sharing of electrons in covalent bonds. The following is an example of a satisfactory description:

> [textbook explains the covalent bond and then gives an example of H_2 molecule] … one hydrogen atom has only one electron that is symmetrically distributed around the nucleus in a 1 s orbital. When two hydrogen atoms form a covalent bond, two atomic orbitals overlap in such a way that the electron clouds are in the region between the two nuclei, and there is an increased probability of finding an electron in this region. According to Pauli's Exclusion Principle, the two electrons of the bond must have opposite spins. (Bekaroğlu and Tan 1986, pp. 74–75)

Three textbooks made a simple mention (M) of Criterion 4 and the following was considered to be an example:

> … a single covalent bond consists of a pair of electrons, with opposite spin, shared by two atoms…. (Ünal 1992, p. 42)

Very few textbooks presented the development of the origin of the covalent bond within a historical perspective. Similar results have been reported for general chemistry textbooks published in the USA (Niaz 2001c). A major finding of the study is that most of the general chemistry textbooks published in Turkey follow an inductivist interpretation of the origin of the covalent bond, which construes Pauli's exclusion principle as the theoretical explanation and ignores the fact that Lewis's cubic atom was crucial for his later explanation of the sharing of electrons. Thus, scientific progress is characterized by a series of theories or models (plausible explanations, from cubic atom to Pauli's principle), which vary in the degree to which they explain the experimental findings. In other words, science does not necessarily progress from experimental findings to scientific laws to theoretical explanations. According to Lakatos (1970, p. 129), the conflict is not between theories and laws but rather between an interpretative and an explanatory theory. Blanco and Niaz (1997) found that many chemistry teachers and students consider progress in science to be characterized by a "Baconian inductive ascent," that is, experimental findings → scientific laws → theoretical explanations. An alternative approach in the present case would be a textbook presentation emphasizing the origin of the covalent bond as a product of conflicting or rival theories (models) for the explanation of bond formation. This shows that appropriate historical reconstructions can benefit students both by providing them with models for alternative/rival approaches and by facilitating a deeper conceptual understanding of the topic.

Oil Drop Experiment

Most chemistry and physics textbooks consider the oil drop experiment to be a simple, classic, and beautiful experiment, in which Robert A. Millikan (1868–1953) by an exact experimental technique determined the elementary electrical charge. The experiment itself has been accepted enthusiastically in many circles without much critical scrutiny. In a poll conducted for *Physics World*, its readers considered the oil drop experiment to be one of ten "most beautiful" of all time (Crease 2002). Furthermore, according to Crease, many respondents considered that the experiment was conceived, carried out, and understood with considerable ease. A historical reconstruction of the events that led to the determination of the elementary electrical charge (*e*) shows the controversial nature of the oil drop experiment then (1910–1925) and that the experiment is difficult to perform even today (Jones 1995).

Holton (1978a) has demonstrated how R.A. Millikan (1868–1953) and F. Ehrenhaft (1879–1952) obtained similar experimental observations, and yet their conceptual frameworks (guiding assumptions) led them to postulate the elementary electrical charge (electrons) and fractional charges (sub-electrons), respectively. It is essential to emphasize that Millikan and Ehrenhaft approached the same experimental data with entirely different guiding assumptions. Modern philosophers of science have emphasized the importance of such assumptions in the progress of science. For example, Holton (1978a) refers to these as presuppositions, Lakatos (1970) as "hard core" of beliefs, and Laudan as guiding assumptions (Laudan et al. 1988). The Millikan–Ehrenhaft controversy lasted for many years (1910–1923) and was discussed at scientific meetings by leading scientists, and both Millikan and Ehrenhaft published original results, critiques, and rebuttals of each other in leading journals (for details, see Niaz 2005). Interestingly, however, after almost 100 years, historians and philosophers of science still disagree on their interpretations of the oil drop experiments (for details, see Niaz 2009).

The Oil Drop Experiment in General Chemistry Textbooks

Presentation of the oil drop experiment in general chemistry textbooks published in USA has been evaluated by Niaz (2000a). Based on a history and philosophy of science (HPS) framework, it was found that very few textbooks dealt with the six criteria satisfactorily (for details, see Chap. 3). Based on the same criteria, Rodríguez and Niaz (2004) analyzed general physics textbooks published in the USA and again found that very few textbooks dealt with the six HPS-based criteria satisfactorily. These studies showed that besides other aspects of the dynamics of scientific progress, the Millikan–Ehrenhaft controversy can open a new window for students, demonstrating how two well-trained scientists can interpret the same set of data in two different ways.

The following criteria based on a historical reconstruction (Niaz 2000a) were used to analyze general chemistry textbooks published in Turkey (Niaz and Coştu 2013):

Criterion 1. Millikan–Ehrenhaft controversy. Millikan and Ehrenhaft obtained similar experimental results, and yet the two interpreted their findings within different theoretical frameworks (guiding assumptions). The controversy started in 1910 with Millikan's critique of Ehrenhaft's method. The controversy turned into a bitter dispute for the next 15 years. According to Millikan, there existed an elementary electrical charge, and charges on all droplets were integral multiples of this fundamental charge. Ehrenhaft argued that the charges on the droplets varied widely, and hence the existence of an elementary electrical charge could not be sustained. This criterion is based on Dirac (1977), Ehrenhaft (1910, 1914), Holton (1978a), and Millikan (1913, 1917).

Criterion 2. Millikan's guiding assumption. Drawing inspiration from Franklin, Faraday, Stoney, Thomson, and others, Millikan formulated the guiding assumption of his research program early in his career. According to this guiding assumption, based on the atomic nature of electricity, Millikan hypothesized the existence of an elementary electrical charge. In his experiments, Millikan found droplets with a wide range of electrical charges. Despite such anomalous data, if it were not for the guiding assumption, Millikan would have abandoned the search for the elementary electrical charge. This criterion is based on Holton (1978a, b) and Millikan (1913, 1917).

Criterion 3. Suspension of disbelief. An important characteristic of Millikan's methodology was to hold the falsification of his guiding assumption in abeyance—that is, suspension of disbelief. In contrast to the traditional scientific method inculcated in school science, Millikan's methodology has found support in modern philosophy of science. This criterion is based on Holton (1978a, b), Lakatos (1970), and Millikan (1913, 1917).

Criterion 4. Transfer of charge as an integral multiple of the elementary electrical charge. Millikan did not measure the charge on the electron itself but rather the transfer of charge on droplets as an integral multiple of the elementary electrical charge (*e*). This criterion is based on Holton (1978a, b) and Millikan (1917).

Criterion 5. Dependence of the elementary electrical charge on experimental variables. The oil drop experiment is extremely difficult to handle. Millikan was constantly trying to improve his experimental conditions to obtain the charge on the droplets as an integral multiple of the elementary electrical charge. Some of the variables that he constantly referred to were evaporation, sphericity, and radius of the droplets, change in density of the droplets, changes in battery voltages, temperature, and viscosity of the air. The oil drop experiment is still difficult to perform in the laboratory. A comparison of Millikan's laboratory notebooks and published results showed that given the complexity of the experimental conditions, he discarded droplets that did not have velocities within a certain range. This criterion is based on Holton (1978a, b) and Millikan (1913, 1965).

Criterion 6. Millikan's experiments as part of a progressive sequence of heuristic principles. Millikan's work started by repeating and a critical evaluation of the experimental work of Townsend, Thomson, and Wilson on charged clouds of water droplets. The first progressive transition was the balanced drop method by using a sufficiently strong electrical field, which later led to the oil drop experiment. It can be argued that Millikan did not design the experiment but rather discovered it. The experiment could have been performed (design) without alluding to the electron theory. Actually, it was the electron theory which suggested the existence of the elementary electrical charge and hence the need for its experimental determination (cf. Holton 1978a, pp. 184–185). This criterion is based on Holton (1978a) and Millikan (1913, 1917, 1950).

For a complete list of Turkish textbooks analyzed in this study, see Appendix 2. Table 7.4 shows that none of the textbooks had a Satisfactory (S) presentation on any of the six criteria. Similar to general chemistry textbooks published in the USA (Niaz 2000a), Turkish chemistry textbooks do not seem to appreciate the importance of controversy in scientific progress (Criterion 1) and therefore deprive students of an opportunity to see how scientists really work.

None of the textbooks described satisfactorily (S) Millikan's guiding assumptions (Criterion 2). Only one textbook made a simple mention (M) of Millikan's guiding assumption in the following terms:

> …Millikan measured charges on the charged droplets [to obtain the charge on the electron]. In the experiments, charges of the droplets were found to be, q = a. 1,6 10^{-19} C. In this equation, a = 1, 2, 3,.. and so forth were integer numbers. These numbers showed that there is no charge lower than 1,6 10^{-19} C on the droplets. [Thus] Millikan assumed that the charge on the electron has to be 1,6 10^{-19} C… (Yavuz 1978, pp. 5–6)

Twenty-six textbooks made no mention (N) of Millikan's guiding assumption and the following is an example:

> In 1909 R. A. Millikan measured successfully both the charge and mass of an electron by performing an experiment known as oil drop experiment (see Fig…) (Hazer 1997, p. 22)

The difference between the two types of presentations can easily be appreciated by a teacher. A presentation classified as Mention (M) attempts to provide some

Table 7.4 Distribution of Turkish general chemistry textbooks according to criteria (oil drop experiment) and classification ($n=27$)

Classification[a]			
Criteria	N	M	S
1	27	–	–
2	26	1	–
3	27	–	–
4	19	8	–
5	27	–	–
6	27	–	–

[a]*N* no mention of the criterion, *M* mention with no details, *S* satisfactory

background reasons (of course, it could have been better) and thus convince the students. On the other hand, the presentation classified as no mention (N) is simply prescriptive, and the student has simply to memorize it. Such presentations can easily be interpreted as an inductive generalization. In other words, the experimental results led Millikan to postulate the elementary electrical charge, and his guiding assumptions played no part.

None of the textbooks mentioned (N) one of the most important features of Millikan's methodology (Criterion 3), that is, in the face of anomalous data, a scientist perseveres with his guiding assumption, holding its falsification in abeyance—in other words, suspension of disbelief. A brief introduction to the historical details would help students to understand how Millikan handled data from his experiments. This could provide an insight for students with respect to how creative imagination of the scientist plays a crucial role.

As seen from Table 7.4, eight textbooks mentioned (M) that Millikan did not measure the charge of thc clcctron itsclf (Critcrion 4) but rather the transfer of charge on droplets as an integral multiple of the elementary electrical charge (e), and the following is an example:

> …after Millikan's many repetitions of the oil drop experiment, he observed that a droplet would gain or lose a charge of an integer multiple of $1.6 \cdot 10^{-19}$ C. From the experiments, he concluded that charges on an oil droplet stem from gaining or losing of one or more electrons. Thus, [he assumed that] the elementary charge of the electron is $1.6 \cdot 10^{-19}$ C (Soydan and Saraç 1998, p. 56)

In contrast to this presentation, consider the following example from a textbook that was classified as no mention (N):

> In 1906, Millikan had calculated charge of the electron and discovered it as $1.6 \cdot 10^{-19}$ C. (Alpaydın and Şimşek 2006, p. 38)

In contrast to Turkish textbooks, it is interesting to note that none of the general chemistry textbooks published in the USA (Niaz 2000a) mentioned (M) that Millikan measured the transfer of charge on droplets as an integral multiple of the elementary electrical charge.

None of the textbooks described satisfactorily (S) or mentioned (M) the different experimental variables that made the oil drop experiment so difficult and its interpretations controversial (Criterion 5). However, three textbooks (Erdik and Sarıkaya 1991; Ergül 2006; Özcan 1998) hinted at the difficulties involved in the experiment, and the following is an example:

> … instead of measuring radius of the droplets, Millikan chose the less erroneous and indirect method to measure [charges of the droplets]. For this purpose, [he] observed the rate of fall of the drop, shortly after it reached limiting velocity because of friction of the air and gravitational force [provides mathematical details]…. (Erdik and Sarıkaya 1991, p. 39)

Interestingly, most textbooks emphasize the experimental nature of chemistry and still ignored how the manipulation of the experimental variables played an important role in the Millikan–Ehrenhaft controversy. None of the textbooks presented (N) Millikan's work as part of a sequence of heuristic principles (Criterion 6). However, it would be helpful if textbooks briefly reviewed some of the earlier experiments that attempted to determine the elementary electrical charge in order to understand the genesis of the oil drop experiment. Before Millikan, Townsend, Thomson and Wilson studied charged clouds of water droplets, which led to the balancing of individual droplets by Millikan and Ehrenhaft and finally Millikan's oil drop experiment. Such an approach can facilitate students' understanding of how scientific endeavor is not a solitary activity but rather a continuous process of critical appraisals (Niaz 2009).

Presentation of the oil drop experiment within a history and philosophy of science (HPS) framework can facilitate understanding of various aspects of nature of science (NOS):

1. Theory-driven nature of scientific observations. Millikan strongly believed in the atomic nature of electricity and did not abandon his theoretical framework (guiding assumptions) in the face of anomalous data. In contrast, Ehrenhaft also had a theoretical framework, viz., nonatomic nature of electricity and hence hypothesized the existence of sub-electrons. Holton (1978a) has suggested that every novice be taught that, "… the graveyard of science is littered with those who did not suspend belief while the data were pouring in" (p. 212). Actually, the suspension of disbelief precisely helps to keep the falsification of guiding assumptions in abeyance.
2. Relationship between scientific constructs and reality. This is a complex issue and is well illustrated by the Millikan–Ehrenhaft controversy. Millikan hypothesized the construct of "electron" in order to understand his experimental observations (reality). In contrast, Ehrenhaft hypothesized the construct of "sub-electrons." The controversy lasted for many years (1910–1923), and there was no instant rationality (conflict resolution), despite the participation of the scientific community.
3. Myth of a stepwise "scientific method." In a sense, by letting his theory to be strictly dictated by his experimental observations, Ehrenhaft was following the scientific method, and still the scientific community in the long run did not support him. Interestingly, Ehrenhaft's theoretical framework was in conformity

with his finding that a series of particles (sub-electrons) with different charges existed. In contrast, Millikan discarded data from many oil drops (as revealed by Holton's, 1978a, inspection of his handwritten notebooks), as he considered these to be not in conformity with his theoretical framework. More recently, Holton (2014b) has clarified Millikan's experimental strategy further by pointing out: "So even if Millikan had included *all* drops and yet had come out with the same result, the error bar of Millikan's final result would not have been remarkably small, but large—the very thing Millikan did not like." Most science teachers and textbooks follow the scientific method and hence would question Millikan's research methodology.

The study reported here can help to improve general chemistry textbooks published in Turkey and other countries. It could also help in the design and implementation of studies that could use HPS-related material to facilitate students and teachers' understanding of the oil drop experiment and consequently the nature of science.

Teaching the Oil Drop Experiment in Lab/Classroom

In recent years, various studies have attempted to improve the presentation of the oil drop experiment by making it more meaningful to the students (cf. Heering and Klassen 2010, 2011, Klassen 2009). Klassen (2009) used an HPS perspective to introduce the oil drop experiment to Canadian undergraduate students, in which the role played by Harvey Fletcher, a student of Millikan, has been emphasized. Fletcher collaborated with Millikan to improve the apparatus and the technique that finally led to the determination of the elementary electrical charge. Millikan claimed that the idea of using oil in the experiment first occurred to him. However, it is possible that the use of oil in the experiment may first have been suggested by Fletcher (1982). Furthermore, the crucial paper (that helped Millikan to win the Nobel Prize) relating to the experiment was published with Millikan as the sole author, whereas the actual work was done jointly. According to Klassen, the inclusion of such details in the classroom can arouse students' interest by showing the human aspect of science through the description of the contributions of both Millikan and his graduate student Fletcher. In this laboratory teaching strategy, students were not left simply to deduce the fact that their data choice is guided by their presuppositions (quantization of electric charge) but rather directed with appropriate questions that challenged them to be reflective about Millikan's work. With this helpful background and guidance, students attained insight and better understanding of the underlying issues, as can be seen from the following response from one of the students:

> By having a preconceived notion of what *e* [charge of the electron] should be, we knew what to expect and disregarded observations that were not expected. In saying so, I believe Millikan depended on his preconceived notion as much as we did. It is likely that when Millikan noticed a quantization trend of the charges, he selected only those drops

> that would illustrate the phenomena and excluded those few that distort it. By doing so, he was able to illustrate his discoveries for us to understand undoubtedly. Had he not, who knows when we would finally acknowledge charge quantization? (p. 604, Reproduced in Klassen 2009)

According to Klassen, the conflict between Millikan and Ehrenhaft provides a unique opportunity to highlight the complexity of the nature of science, namely, that Ehrenhaft's actions were guided by the traditional scientific method, whereas Millikan's actions were guided by his presuppositions about the quantization of the electrical charge. In a similar vein, Kolstø (2008) has referred to yet another aspect of nature of science, namely, the social milieu in which the controversy developed:

> Nevertheless, the different results of Millikan and Ehrenhaft forced the scientific community to argue and debate the adequacy of the different theories, methodologies, interpretations and results. Also in their papers, Millikan and Ehrenhaft put forward criticisms of each other's methods and results. The Millikan story therefore indicates that narratives from the history of science might be used to illustrate the important role argumentation and criticism play in scientific knowledge production. *It was not Millikan or Ehrenhaft who decided that Millikan's knowledge claim was correct, but the scientific community they were part of.* (p. 987, emphasis added)

Interestingly, the same historical episode (oil drop experiment) led Klassen (2009) to recognize the importance of presuppositions vis-à-vis scientific method, and Kolstø (2008) the social and historical milieu of the two protagonists. Indeed, this is a manifestation of how two different NOS aspects can emerge from the same domain-specific historical episode. The role played by the scientific community in the Millikan–Ehrenhaft controversy was important and is generally ignored in science textbooks. The recently opened Nobel Prize Archives show that although Millikan was nominated for the prize from 1916 onwards, it was recommended that the prize be withheld as the controversy with Ehrenhaft was not yet resolved (Holton 1988). Millikan finally got the Nobel Prize in 1923. This clearly shows that given the opportunity to think, reflect, interact, and discuss, the historical background of the experiment (domain-specific) can provide students a better picture of the scientific enterprise and facilitate a better understanding of the nature of science.

In the context of the oil drop experiment, Paraskevopoulou and Koliopoulos (2011) designed a study for high school students in Greece, in order to facilitate an understanding of the nature of science based on the Millikan–Ehrenhaft dispute. Based on a historical reconstruction of the oil drop experiment (Holton, 1978a; Niaz 2000a), they prepared short stories for students that were discussed in class, as suggested by Clough and Olson (2004) within an explicit and reflective framework (Akerson et al. 2000). As an example of this conceptual framework, students were asked the following question: How was it possible for these two scientists (Millikan and Ehrenhaft) to reach different conclusions when they were looking at the same data? (p. 958). Based on this experience with the students, the authors report evidence for improvement in students' understanding of the following NOS aspects and how these were emphasized during class discussions: (a) The role played by empirical data in scientific debate. The existence of empirical data is essential for critical evaluation and the debate to begin. Both Millikan and Ehrenhaft attempted to improve their experimental methods to remove possible errors. (b) The distinction between observation and inference. Millikan's neglected observations (based on Holton 1978a) and the conclusions of his inferences. How while discussing the same observations can lead to two different inferences? (c) The role of the scientist's imagination and creativity in the formation of theory. (d) The natural sciences have a subjective content during the formation of a theory. The authors elaborated on this aspect: "The 'battle over the electron' demonstrated that the idea that the charge is quantized does not derive from the empirical data itself but from its interpretation, which is based on the subjective hypothetical knowledge adopted by each researcher. Of course, this element of NOS if presented in the classroom one-sidedly could lead to relativist epistemological concepts, which are not part of the aims of the proposed teaching intervention" (p. 947). It is important to note that all four NOS aspects studied by the authors are directly based in the context of the oil drop experiment. This clearly shows the importance of how domain-specific NOS strategies can be developed and provides students an opportunity to understand the dynamics of scientific progress and science in the making (cf. Niaz 2009, 2012a). Furthermore, the authors draw attention to the fact that this teaching strategy can easily be implemented in the classroom without changing the curriculum as the basic information about the oil drop experiment is already present in the textbooks. Once again, this leads to the thesis that the history of chemistry is already "inside" chemistry and what we need to do is a historical reconstruction of the different episodes or experiments (Niaz and Rodríguez 2001).

Electrolyte Solution Chemistry

It is generally considered that 1885 was the "foundation date" of physical chemistry, when Wilhelm Ostwald (1853–1932), Jacobus van't Hoff (1852–1911), and Svante Arrhenius (1859–1927) started the first journal related to physical chemistry, *Zeitschrift für Physikalische Chemie* (Gavroglu 2000). Actually, physical chemistry went through a long period of gestation starting in the 1880s and lasting until the early 1930s. This period is of interest to chemistry educators as it shows how chemists worked during the nascent years, reasoned, and argued among themselves, as controversies arose. An example of this work in physical chemistry is provided by van't Hoff who formulated a theory of osmosis based on theoretical considerations, which was controversial (De Berg 2014b). The controversy raged between the European school (referred to as the ionist school) of Svante Arrhenius, Wilhelm Ostwald, and Jacobus van't Hoff, who believed that salts partially dissociated when dissolved in water, and the British school (referred to as the hydrationist school) of Henry Armstrong (1848–1937), Spencer Pickering (1858–1920), and George Fitzgerald (1851–1901) who regarded the dissociation hypothesis as unthinkable and lacking in firm laboratory evidence. A variety of experimental data were used by the European school to promote the idea of the dissociation of salts in aqueous solution. Measurements of electrical conductivity, boiling point elevation, freezing point depression, vapor pressure lowering, osmotic pressure, and heats of neutralization were among the techniques of interest. There were, however, dual (alternative) interpretations of the data. The European school insisted that the measurement of the osmotic pressure of a range of aqueous salt solutions was best interpreted in terms of the partial dissociation of the dissolved substance. On the other hand, the British school contended that as laboratory experience confirmed the production of hydrated compounds from solution, it provided evidence that solution was an association with water phenomenon rather than a dissociation in water phenomenon.

So while the European school was classified as the ionist or dissociationist school, the British school was classified as the hydrationist or associationist school. Throughout the late nineteenth century and early twentieth century, the ionists insisted on explaining osmotic pressure by focusing on the solute, whereas hydrationists explained osmotic pressure by focusing on the solvent.

Armstrong (1928) a strong supporter of the hydrationist school expressed his arguments in a somewhat picturesque term: "… the physical chemist has been neither chemist nor physicist at heart. The mutation from chemist to physical chemist certainly seems to have involved the loss of the primary factor in chemistry: chemical feeling" (p. 51). When W. L. Bragg published the results of his X-ray study of sodium chloride and concluded that no molecules of sodium chloride (NaCl) existed as such but rather sodium and chloride ions were distributed in a chessboard fashion in a three-dimensional lattice, Armstrong (1927) protested that this model:

> … is repugnant to common sense, absurd to the nth degree, not chemical cricket. Chemistry is neither chess nor geometry whatever X-ray physics may be. Such unjustified aspersion of the molecular character of our most necessary condiment must not be allowed any longer to pass unchallenged. A little study of the Apostle Paul may be recommended to Professor Bragg as a necessary preliminary even to X-ray work,. . ., that science is the pursuit of truth. It were time that chemists took charge of chemistry once more and protected neophytes against the worship of false gods: at least taught them to ask for something more than chessboard evidence." (p. 478)

This controversy and the arguments (chemical feeling, chemical cricket, pursuit of truth, false gods) put forward in prestigious journals may seem curious and quaint to some of our students. Nevertheless, these do form an important part of the "science in the making" and understanding NOS in a particular context (Niaz 2012a; also see Chap. 3 for this controversy within a NOS context). At this stage it is important to consider the following dilemma raised by De Berg (2014b): Is a generalized broad picture of NOS any different to that broad generalized notion of the "scientific method" which has received significant criticism from historians and philosophers of science? Furthermore, could the so-called myth of the scientific method be replaced with an equally spurious NOS? Indeed, this is an issue that requires discussion and consensus within the science education community. Again, De Berg (2014) has insisted: "As far as tertiary level science education is concerned, there are some distinct advantages (one of which is authenticity) in uncovering NOS issues in specific science content" (p. 6). One could agree with respect to the "authenticity" of scientific practice. However, how do we justify restricting this to tertiary level science education? In my opinion, both at the high school and introductory university level, it is possible to integrate domain-specific issues (e.g., electrolyte solution chemistry) based on historical reconstructions and then integrate this experience within a domain-general framework of NOS. Pedagogically this makes sense as students may encounter similar episodes in their physics and biology courses, which may help them to understand the scientific endeavor in a wider context.

Teaching Electrolyte Solution Chemistry in the Classroom

De Berg (2014) has suggested the inclusion of the controversy with respect to electrolyte solution chemistry in the classroom by emphasizing the dual interpretation of experimental data within a domain-specific context:

> … rich historical data on solution chemistry is available for constructing a class debate between an aspiring Armstrong's team and a daring Arrhenius' team, or alternatively, the construction of a critical assignment or a sample of Interactive Historical Vignettes. It has also been demonstrated that recent developments in physical chemistry have been drawing upon some of the historical elements of the electrolyte solution debate to construct a more explanatory model of solution behaviour than that relying on empirical coefficients based on a model of complete dissociation for strong electrolytes." (p. 10; for details, see Heyrovska 1996)

It is suggested that the students are provided with the following description and relevant data, and the ensuing questions then can be debated. In the late nineteenth century, ionists explained osmotic pressure by focusing on the solute, whereas hydrationists explained osmotic pressure by focusing on the solvent (De Berg 2014b, p. 10):

(a) Hydrationists said that an increase in osmotic pressure was caused by an increase in the number of free water molecules that become bound to the salt. Does your data agree with this proposition? Explicitly illustrate using the data.
(b) Ionists said that an increase in osmotic pressure was caused by enhanced dissociation of the salt in water. Does your data agree with this proposition? Explicitly illustrate using the data.

The Photoelectric Effect

The photoelectric effect constitutes an important part of the science curriculum and is the usual starting point for the introduction of quantum theory to undergraduates. Throughout the latter part of the nineteenth century, light was considered to be a wave propagating in an all-pervading medium. Properties such as diffraction, interference, and polarization convinced physicists that visible monochromatic light is a periodic transverse oscillation—known as the classical wave theory of light. Based on this theory, however, scientists had difficulties in explaining the photoelectric effect, namely, the observation that many metals emit electrons when light shines upon them. The classical wave theory of light attributed this effect to the transfer of energy from the light to an electron in the metal and eventually dislodging the electron. However, experiments showed that the light used to dislodge electrons in the photoelectric effect had a *threshold frequency*, below which no electrons were emitted. In 1905, Einstein attempted to solve this anomaly by suggesting that light behaves as though it consists of a stream of independent localized units of energy that he called *light quanta*. Consequently, an electron in an atom will receive energy from only one light quantum at a time, and this helped to explain the role of threshold frequency.

The Photoelectric Effect in Textbooks

A historical reconstruction of the events that culminated in Einstein's hypothesis of light quanta to explain the photoelectric effect and the ensuing controversy in the scientific community was first elaborated by Niaz (2009, Chap. 8). Subsequently, Niaz et al.(2010a) elaborated the following six criteria for evaluating science textbooks:

1. *Lenard's trigger hypothesis to explain the photoelectric effect.* Lenard made the important experimental determination that in photoelectric phenomena the velocity of the ejected electrons is independent of the intensity of light. According to his trigger hypothesis, electrons in an atom already have the necessary potential energy, and the incident light only triggers the release of the selected electrons. This criterion is based on Lenard (1902) and Wheaton (1983). For this criterion to be met, it is important for the textbooks to describe the following aspects:

   (a) Lenard in 1902 strongly believed in the wave theory of light.
   (b) Velocity of the ejected electrons was independent of the intensity of light.
   (c) Electrons in an atom already had the necessary potential energy.
   (d) Incident light only triggers the release of the selected electrons.

2. *Einstein's quantum hypothesis to explain the photoelectric effect.* According to Einstein, if light consists of localized quanta of energy, an electron in an atom will receive energy from only one light quantum at a time. Based on this hypothesis, Einstein predicted that the stopping potential, when plotted against the frequency of the incident light, would give a straight line whose slope would provide Planck's constant *h*. Furthermore, Einstein's hypothesis constituted a rival explanation to Lenard's triggering hypothesis to explain the photoelectric effect. This criterion is based on Einstein (1905), Holton (1999), and Wheaton (1983), and it is important for the textbooks to describe the following aspects:

   (a) Einstein's quantum hypothesis constituted a rival to Lenard's trigger hypothesis.

(b) Einstein explained the finding that the velocity of ejected electrons would depend on the frequency and not the intensity of incident light.
(c) Light consists of localized quanta of energy, so an electron in an atom will receive energy from only one light quantum at a time.
(d) Einstein predicted that the stopping potential of the metal when plotted against the frequency of the incident light would give a straight line, whose slope would provide Planck's constant *h*.

3. *Lack of acceptance of Einstein's quantum hypothesis in the scientific community.* Although Einstein presented his interpretation of the photoelectric effect based on the quantum hypothesis in 1905, it was generally rejected by the scientific community. The main objection against Einstein's hypothesis was that it seemed to refute the highly accepted, classical wave theory of light. Even Planck, the "originator" of the quantum theory, opposed Einstein's hypothesis until about 1913. It took many years for this "revolutionary theory" to be accepted in the body of scientific knowledge. This criterion is based on Einstein (1905) and Wheaton (1983). For this criterion to be met, it is important for the textbooks to describe the following aspects:

   (a) Truly novel ideas are generally accepted very slowly.
   (b) Einstein's hypothesis was not accepted by the scientific community, including Planck the "originator" of the quantum hypothesis, for many years.
   (c) The main objection to Einstein's hypothesis was that it seemed to refute the highly accepted classical wave theory of light.

4. *Millikan's experimental determination of the Einstein photoelectric equation and Planck's constant h.* Millikan provided the first direct experimental proof of the exact validity of the Einstein equation ($\frac{1}{2} mv^2 = Pe = h\nu - p$) and the first direct photoelectric determination of Planck's constant *h*. According to Holton (1999), "Ironically, it had been Millikan's experiment [Millikan 1916] which convinced the experimentalist-inclined committee in Stockholm to admit Einstein to that select circle [Nobel Prize]" (p. 235). For this criterion to be met, it is important for the textbooks to describe the following aspects:

   (a) Underdetermination of scientific theories by experimental evidence, viz., no amount of experimental evidence can provide conclusive proof for a theory.
   (b) Experimental details of Millikan's determination of Einstein's photoelectric equation and Planck's constant *h*.
   (c) The graph of stopping potential against frequency, whose slope would provide Planck's constant *h*.

5. *Millikan's presuppositions about the nature of light.* Although Millikan provided the first experimental proof of Einstein's equation, he considered Einstein's interpretation of the photoelectric effect based on the quantum hypothesis as "the reckless, hypothesis" (Millikan 1916, p. 355). Millikan's opposition to the quantum hypothesis is attributed to his prior presupposition and a strong belief in the classical wave theory of light. This criterion is based on Holton (1999),

Millikan (1916), and Wheaton (1983). For this criterion to be met, it is important for the textbooks to describe the following aspects:

(a) Before doing an experiment, scientists invariably do have prior theoretical beliefs or presuppositions and they resist any change in those epistemological beliefs.
(b) In the present case, Millikan strongly believed in the wave theory of light.
(c) Millikan (1916) presented experimental evidence to support Einstein's photoelectric equation and in the same paper considered his underlying hypothesis to be "... the bold, not to say the reckless hypothesis ..." (p. 355).

6. *The historical record presented and its interpretation within a history and philosophy of science perspective.* A historical reconstruction of the photoelectric effect shows that it is based on a series of experimental findings intertwined with their interpretations based on different theoretical frameworks. In order to facilitate a better understanding of the photoelectric effect, it is essential that textbooks are consistent in attributing the different experimental findings and their interpretations to the relevant scientists. For example, Millikan accepted Einstein's equation but not his interpretation. Textbooks tend to confound the issues by attributing to Millikan the acceptance of both. This is strictly based on the historical record and may give extra points to textbooks that do represent the historical part well. Most textbooks would perhaps mention Einstein's hypothesis and Millikan's experimental determination; thus, to classify as Mention (M), textbooks need to make an extra effort of including one more aspect. It is important for the textbooks to describe the following aspects:

   (a) Lenard's experimental findings and his trigger hypothesis
   (b) Einstein's hypothesis to explain the photoelectric effect
   (c) Opposition to the acceptance of Einstein's hypothesis in the scientific community
   (d) Millikan's experimental determination to provide evidence for Einstein's equation and the determination of *h*
   (e) Millikan's presuppositions that led him to reject Einstein's quantum hypothesis

Depending on the elaboration of the different aspects, on each of the six criteria, textbooks were classified as Excellent (E), Satisfactory (S), Mention (M), or No mention (N). Niaz et al. (2010a) analyzed 103 general physics textbooks published in the USA (ranging from the 1950s to the 2000s). For most textbook authors, the historical aspect seems not to have been part of an overall writing strategy or presentation. Results obtained showed that 3 % of the textbooks were classified as Excellent (E) on Criterion 3. None of the textbooks were classified as excellent on the other five criteria. Percentage of general physics textbooks that were classified as No mention (N) on the six criteria were the following: Criterion 1 = 84 %, Criterion 2 = 14 %, Criterion 3 = 84 %, Criterion 4 = 69 %, Criterion 5 = 95 %, and

Criterion 6 = 91 %. It seems that most of the historical aspects are not only ignored by general physics textbooks but also distorted.

Based on the same six criteria as presented above, Ospina (2010) has analyzed the presentation of the photoelectric effect in 118 general chemistry textbooks published (ranging from the 1950s to 2000s) in the USA. Overall, general chemistry textbooks include even less historical details than general physics textbooks, and the results are quite similar. The percentage of general chemistry textbooks that were classified as No mention (N) were the following: Criterion 1 = 100 %, Criterion 2 = 48 %, Criterion 3 = 92 %, Criterion 4 = 98 %, Criterion 5 = 100 %, and Criterion 6 = 100 %. The following is an example of a general chemistry textbook presentation that was classified as Satisfactory (S) on Criterion 2:

> A second paradox arose from the discovery of the *photoelectric effect.* In this effect, a beam of light falling on a metal surface in an evacuated space ejects electrons from the surface, causing an electric current (called a *photocurrent*) to flow. This phenomenon was in itself not difficult to understand, because electromagnetic radiation carries energy. The problem came in explaining the observed dependence of the effect on the frequency of the light. It was found that for light frequencies less than a certain threshold frequency ν_0, no electrons were ejected; once the frequency exceeded the threshold, the photocurrent increased rapidly. According to classical theory, the energy associated with electromagnetic radiation depends only on the intensity (or square of the magnitude of the electric field), and not on the frequency. Why, then, could a very weak beam of blue light eject electrons from sodium when an intense red beam had no effect (Fig. 17-8a)? Further experiments measured the kinetic energy of the ejected electrons and revealed a linear dependence of the maximum kinetic energy on frequency as shown in Figure 17-8b … This behavior also was inexplicable in classical physics. (Oxtoby et al. 1990, pp. 674–675, emphasis and italics in original)

At this stage, it would be interesting to ask: Why was this textbook classified as Satisfactory (S) and not Excellent (E)? Some of the positive aspects of this presentation are the following: (a) observed dependence of the photoelectric effect on the frequency of the light, (b) failure of the classical theory to explain the effect and hence the importance of Einstein's hypothesis of *light quanta* (it also considers the photoelectric effect as a paradox), and (c) linear dependence of the maximum kinetic energy on frequency of light and not intensity. Interestingly, these three aspects are precisely those that were included for the evaluation of Criterion 2. The only aspect missing is the one related to the following: Einstein's quantum hypothesis constituted a rival to Lenard's trigger hypothesis (which was the subject of Criterion 1). This clearly shows what constitutes a historical reconstruction. This textbook complies with three aspects and ignores the fourth, namely, the reference to Lenard's rival hypothesis. This rivalry lasted for many years and was even supported by Max Planck, who suggested the quantum analog of the triggering hypothesis (Wheaton 1983, p. 179). Consequently, if this textbook had referred to Lenard's trigger hypothesis, it could have been classified as Excellent (E). This also shows that the history of science (physics or chemistry) is already inside the subject matter, and what we need to do is "weave" different elements of "science in the making" into a coherent story (cf. Niaz 2012a).

Another example is provided from a general chemistry textbook that was classified as Satisfactory (S) on Criterion 3:

> Einstein's theory with respect to light was a dilemma for the scientists. On the one hand, the theory explains satisfactorily the photoelectric effect. But, on the other hand the theory of the particle of light was not consistent with its known behavior as a wave. The only way to resolve this dilemma was to accept the idea that light possessed properties both of the particle and wave. Depending on the type of experiment, light behaves as a wave or torrent of particles. This concept differed radically from what the physicists thought about matter and radiation and took a long time for it to be accepted. (Chang 2007, p. 274)

This is a fairly good description of the dilemma faced by the scientific community with respect to Einstein's hypothesis of light quanta. It could have been classified Excellent (E) if it had referred to another aspect related to NOS, namely: Truly novel ideas are generally accepted very slowly. Again, this illustrates how NOS can be introduced in the classroom within a domain-specific context of the curriculum and that history of science is "inside" science. Interestingly, some textbooks refer to the NOS aspect without referring to the context in which it could be understood. Precisely, an integrated view would lead to integrating both approaches, namely, (a) from a domain-specific context to understanding some general NOS aspect and (b) from a domain-general context to illustrating a particular context.

It is plausible to suggest that the inclusion in science textbooks of the following aspects related to the photoelectric effect can facilitate a better understanding of the dynamics of scientific progress: (a) Einstein's quantum hypothesis constituted a rival to Lenard's trigger hypothesis. (b) Einstein's hypothesis was not accepted by the scientific community, including Planck, the "originator" of the quantum hypothesis, for many years. (c) Millikan presented experimental evidence to support Einstein's photoelectric equation and still rejected his quantum hypothesis. (d) Scientific theories are underdetermined by experimental evidence, that is, no amount of experimental evidence can provide conclusive proof for a theory (cf. Giere 1999, p. 237). (e) Scientists customarily have prior theoretical beliefs or presuppositions before doing an experiment, and they resist any change in those beliefs. (f) An overview of the historical reconstruction as provided in Criterion 6 (Lenard, Einstein, and opposition, Millikan's experiments and presuppositions) can help teachers and textbook authors to coordinate the different objectives. Inclusion of these aspects will help to facilitate an integrated view of NOS and to introduce the historical aspects of the photoelectric effect as an unfolding story.

The Photoelectric Effect in Laboratory Manuals

The photoelectric experiment has the advantage of integrating both the experimental and theoretical aspects in a particular scientific context. Actually, the experiment makes sense only if the underlying issues related to the work of Millikan (experimental) and Einstein (theoretical) are emphasized. It seems that during the period 1940–1950, the experiment was used in the undergraduate laboratory sporadically.

However, starting in 1961, it has become a regular part of the undergraduate laboratory in many parts of the world. Klassen et al. (2012) have analyzed 38 laboratory manuals (electronically published instructions) from different parts of the world. Based on a history and philosophy of science framework (same as presented in Niaz et al. 2010a and summarized above), manuals were analyzed according to the following four criteria:

Criterion 1: Einstein's quantum hypothesis to explain the photoelectric effect
Criterion 2: Lack of acceptance of Einstein's quantum hypothesis in the scientific community
Criterion 3: Millikan's experimental determination of the Einstein photoelectric equation and Planck's constant *h*
Criterion 4: Millikan's presuppositions about the nature of light

All laboratory manuals were classified in a similar way as in Niaz et al. 2010a and presented above: E = Excellent, S = Satisfactory, M = Mention, and N = No mention. Results obtained showed that none of the manuals was classified as Excellent. The percentage of manuals that were classified as No mention were the following: Criterion 1 = 61 %, Criterion 2 = 100 %, Criterion 3 = 92 %, and Criterion 4 = 100 %. It was expected that as Criterion 3 dealt with Millikan's experimental verification of Einstein's quantum hypothesis, most manuals would perhaps fare well. However, only 8 % of the manuals were classified as Mention (M). The following is an example of a presentation that was classified as No mention (N):

> Einstein's model thus predicts two things: that the voltage required to stop the photoelectric effect from occurring should be independent of the intensity of the light, and that we should obtain a linear relation between the stopping potential, V_s, and the frequency, ν, of the light with which we illuminate the photocathode. (Mount Holyoke College 2004, reproduced in Klassen et al. 2012)

Although this presentation described the relationship between Einstein's predictions and the photoelectric effect, it made no connection with Millikan's experimental measurements. In general, just like the textbooks, the manuals ignored the historical context and the difficulties involved in understanding the experimental data that led to alternative interpretations.

The Photoelectric Effect in the Classroom

Based on a history and philosophy of science perspective (HPS), Oh (2011) has designed a study to facilitate Korean freshman students' conceptual understanding of the photoelectric effect. Besides the HPS perspective, the teaching strategy is based on previous experience in science education related to the introduction of cognitive conflict in the classroom (Lee and Yi 2013; Niaz 1998a; Tsai 2000). In this study the cognitive conflict effectively integrated students' existing experiences or knowledge, new theories, anomalous data based on discrepant events, and perceptual supports to facilitate the development of scientific conceptions (Oh 2011, p. 1156). One of the most significant alternative conceptions (misconception) found was the following: Students hold that the longer the time taken by the incident light, the greater the frequency, and the *more energy stored in electrons inside the metal* leading to more active chemical reaction and consequently, the higher the maximum kinetic energy of the ejected electrons (p. 1158; compare the part in italics with Lenard's trigger hypothesis). Indeed, this sounds interesting if we compare it with the historical record. Before Lenard (1902) postulated his trigger hypothesis, scientists generally interpreted the photoelectric effect within the wave theory of light. Even Lenard considered that electrons in an atom already possessed their photoelectric velocity or the potential energy equivalent (Wheaton 1983). In contrast, Einstein (1905) had postulated that light consists of localized quanta of energy and an electron will receive energy from only one light quantum at a time. In other words, even Lenard's hypothesis had elements of classical theory. The tension between the classical and quantum theory has been noted by Oh (2011): "Our study emerged from the hypothesis that the core [negative heuristic, according to Lakatos 1970] of students' alternative conceptions originated in classical physics, which is not consistent with modern physics [quantum theory]" (p. 1158). This is an important finding and shows the similarity between students' alternative conceptions and scientists' concepts in the past, and the two are considered to be commensurable. Finally, Oh (2011) found that those students who did not manifest this alternative conception had a better understanding of other aspects of the photoelectric effect.

The relationship between classical and quantum concepts is complex and at the same time needs exploration. For example, in the context of the photoelectric effect, one electron can be stripped away from a helium atom that is exposed to ultraviolet light below a certain wavelength. This threshold wavelength can be determined experimentally to very high accuracy (Styer 2000). In contrast, classical mechanics predicts that light of any wavelength will strip away an electron. It is plausible to suggest that any conceptual understanding of quantum mechanics will also require a reference to classical ways of thinking (especially at the introductory high school and university level), or in other words quantum mechanics approaches classical mechanics as a limiting case:

> … classical mechanics eventually gave way to the quantum theory, which is very different in its basic structure, but which still contains classical theory as a limiting case, valid approximately in the domain of large quantum numbers. Agreement with experiments in a limited domain and to a limited degree of approximation is evidently no proof, therefore, that the basic concepts of a given theory have a completely universal validity. (Bohm 1980, p. 82)

Transition from classical to quantum ideas is difficult for the students and any example that helps to establish a bridge is helpful. Indeed, this is sound advice if we want to facilitate students' conceptual understanding beyond classical mechanics not only in the photoelectric effect but also wave–particle duality, which is the subject of the next section.

Wave–Particle Duality

In most parts of the world, general chemistry curricula and textbooks present atomic structure by referring to the work of J. Dalton, J.J. Thomson, E. Rutherford, N. Bohr, and A. Sommerfeld. Following this, the photoelectric effect and its interpretation by Einstein is presented as an application of quantum theory. In a sense Bohr's model of the atom departed from the classical wave theory of light by introducing the "quantum of action." Paradoxically, Bohr held on to the wave theory of light until the mid-1920s. Next, in order to introduce the wave mechanical model of the atom (E. Schrödinger), L. de Broglie's contribution is mentioned by posing the question: If light can have both wave and particle properties, then why particles of matter (e.g., electrons) cannot also have both properties? Furthermore, experimental work of C. Davisson and L.H. Germer is reported based on diffraction of electron beams by metal foils. All these developments are presented in a straightforward chronological sequence as if there were no conceptual or other problems associated or difficulties involved. On the contrary, postulation of the wave–particle duality was a controversial topic from the very beginning and is closely enmeshed with the origin and development of the photoelectric effect based on Einstein's hypothesis of light quanta and quantum theory. De Broglie (1924) in a seminal paper did not reject the wave theory of light but rather explored the reconciliation of light quanta with "… the strong experimental evidence on which was based the wave theory" (p. 446).

Wave–Particle Duality in General Chemistry Textbooks

A historical reconstruction of wave–particle duality was first elaborated by Niaz (2009, Chap. 12). Subsequently, Niaz and Marcano (2012) elaborated the following six criteria for evaluating 128 general chemistry textbooks published (1954–2011) in the USA:

Criterion 1: Einstein and de Broglie suggested wave–particle duality before there was any conclusive experimental evidence. The origin of the concept of wave–particle duality can be traced to Einstein's (1905) hypothesis of the light quantum to explain photoelectric effect. This was followed by important theoretical formulations by Einstein (1909a, b, 1916) and de Broglie (1922, 1923a, b, 1924). Despite Einstein's prestige and authority, duality remained a controversial hypothesis, until conclusive experimental evidence was presented by Davisson and Germer (1927a). This clearly shows how theoretical formulations do not necessarily follow from experimental data. For this criterion to be met, it is important for the textbooks to describe the following aspects:

(a) The origin of wave–particle duality can be attributed to Einstein and de Broglie.
(b) Experimental evidence was presented later by Davisson and Germer (1927a).

Criterion 2: De Broglie suggested how matter waves could be observed experimentally. From the very beginning, de Broglie (1923a, b) not only hypothesized that waves were associated with material particles but also suggested how these could be confirmed experimentally by passing beams of electrons through small openings and observing diffraction phenomena. However, it is important to note that initially even some of the leading physicists (Maurice de Broglie, P. Langevin, J. Perrin, A. Dauvillier, W. Elsasser, J. Franck) found either the experiments too difficult or simply not worth the effort (Medicus 1974). For this criterion to be met, it is important for the textbooks to describe the following:

(a) De Broglie not only presented the hypothesis that waves are associated with particles but also suggested that these could be confirmed experimentally.
(b) Some physicists found the experiments difficult or simply not worth the effort.

Criterion 3: Importance of Davisson–Germer experiments and their struggle to interpret experimental data. Early experiments by Davisson (Davisson and Kunsman 1923) were considered by W. Elsasser as providing evidence for wave–particle duality. However, Davisson did not agree and went on to perform further experiments, leading to an accident in the laboratory which provided a clue to the problem. Interpretation of the experimental data was, however, difficult, and as late as August 1926, Davisson was consulting other physicists at a meeting in Oxford (Davisson and Germer 1927a). Finally, in a note to *Nature*, April 1927, Davisson reported: "These results are highly suggestive, of course, of the ideas underlying the theory of wave mechanics" (Davisson and Germer 1927b, p. 558). For further details, see Navarro (2012). It is important for the textbooks to describe the following aspects:

(a) Early experiments did not convince Davisson as providing support for the wave–particle duality, despite support from some physicists.
(b) Later experiments (Davisson and Germer 1927a, b) were also difficult to interpret and required the help of the scientific community.

Criterion 4: Role of experiments by G.P. Thomson. Thomson was also independently working to provide experimental evidence for wave–particle duality and reported his results to *Nature* about two months after Davisson (Medicus 1974; Navarro 2012). This shows how various research groups may work on the same problem and arrive at the same conclusion using different experimental techniques. Furthermore, it illustrates that there is no one way of doing science. For this criterion to be met, it is important for the textbooks to describe the following aspects:

(a) G.P. Thomson provided experimental support for the wave–particle duality independently.
(b) Different groups of scientists can work on the same problem using diverse experimental techniques and arrive at the same conclusion.

Criterion 5: Controversial nature of wave–particle duality and de Broglie's reputation as an obstacle in the acceptance of his theory. When De Broglie first presented his ideas on wave–particle duality, he was a mature scientist and had considerable research experience (having published about two dozen scientific papers on electron, atomic, and X-ray physics, before his doctorate). Despite this he had to face opposition and criticism on two grounds: (a) Wave–particle duality was a controversial issue and required physicists to give up their previous strong belief in the dominant classical wave theory of light. (b) De Broglie's previous research had led him into controversies with two influential schools of physicists (Copenhagen and Munich). This shows how in scientific development, innovative and creative ideas are resisted and even rejected due to the previous reputation of the scientist. For this criterion to be met, it is important for the textbooks to describe the following aspects:

(a) Wave–particle duality was a controversial thesis as it required the physicists to abandon their previous belief.
(b) De Broglie's previous research experience had led him into controversies with two influential schools of physicists.

Criterion 6: Why was it Schrödinger who developed de Broglie's ideas? Despite de Broglie's reputation, Einstein started to support his ideas soon after he received his doctoral thesis from Langevin. This support was crucial in convincing Schrödinger to develop de Broglie's ideas. At first Schrödinger was skeptical; however, later he acknowledged that de Broglie's ideas were a source of inspiration. Given de Broglie's reputation (especially in the Copenhagen school), Schrödinger's own previous interests, expertise in theoretical spectroscopy, and compatibility with Einstein on various problems of quantum mechanics, it was almost natural for him to have developed wave–particle duality (Raman and Forman 1969). It is important for the textbooks to describe the following aspects:

(a) Early support of Einstein for de Broglie's ideas
(b) Schrödinger's acknowledgment that de Broglie's ideas were a source of inspiration

For each criterion, textbooks were expected to describe two aspects (a and b) and based on these were classified as Satisfactory (S) (description of both aspects (a) and (b)), Mention (M) (description of aspect (a) or (b)), and No mention (N) (none of the two aspects). Results obtained showed that none of the textbooks was classified as Satisfactory (S) on criteria 2, 3, 5, and 6. The following were the percentages of textbooks that were classified as Satisfactory (S): Criterion 1 = 20 % and Criterion 4 = 3 %. At this stage it would be interesting to consider two examples of textbook presentations that were classified as Satisfactory (S) on Criterion 1:

> Energy, prior to 1900, was not considered to consist of particles. It was noncorpuscular in nature, and therefore continuous. It was this distinction between matter and energy that had been abandoned by Planck in 1900, by Einstein in 1905, and again by Bohr in 1913 … The French physicist Louis de Broglie proposed in 1924 that not only light but *all* matter has a dual nature and possesses both wave and corpuscular properties. He reasoned that there

> should be symmetry in nature: If a radiant corpuscle—that is, a photon—has a frequency and a wavelength and therefore has wave properties, why should not a material particle also have wave properties? (p. 429, original italics) … When de Broglie first published his wave theory of matter, there was no experimental evidence to support his bold hypothesis. Within three years, however, two different experiments had been performed that demonstrated the diffraction of a beam of electrons. Clinton J. Davisson, assisted by L.H. Germer, … observed the diffraction of electrons when a beam of electrons was directed at a nickel crystal. (Segal 1989, p. 431, underlined added)

> Einstein used the photoelectric effect to demonstrate that light, which is usually thought of as having wave properties, can also be thought about in terms of particles or massless photons. This fact was pondered by Louis Victor de Broglie (1892–1987). If light can be considered as sometimes having wave properties and other times having particle properties, he asked why doesn't matter behave similarly? That is, could a tiny object such as an electron, which we have so far considered a particle, also exhibit wave properties in some experiments? … This idea was revolutionary, since it linked the particle properties of the electron (m and υ) with possible wave properties (λ). Experimental proof was soon produced. Davisson and Germer, … found that a beam of electrons was diffracted like light waves by the atoms of a thin sheet of metal foil and that de Broglie's relation was followed quantitatively … After de Broglie's suggestion that an electron can be described as having wave properties, a great debate raged in physics. How can an electron be described as *both* a particle and a wave? … One can only conclude that *the electron has dual properties*. The result of a given experiment can be described *either* by the physics of waves *or* by the physics of particles; there is no single experiment that can be done to show that the electron behaves *simultaneously* as a wave and a particle! (Kotz and Purcell 1991, italics in the original, underlined added)

Interestingly, the presentation by Segal (1989) is almost a historical reconstruction of wave–particle duality, starting with Planck in 1900 to Einstein in 1905, Bohr in 1913, de Broglie in 1924, and finally three years later Davisson and Germer (1927a). The underlined part is very important as very few textbooks refer to it, namely, de Broglie's conceptualization of wave–particle duality preceded its experimental determination by Davisson and Germer. Kuhn (1978), however, has questioned the role of Planck as an originator of the quantum hypothesis. According to Kuhn, Planck's contribution represented a mathematical or calculation device to understand black-body radiation. On the other hand, the originator of the quantum hypothesis was Einstein, who emphasized the physical significance of the postulate. Niaz and Fernández (2008) have analyzed the role of the origin of the quantum hypothesis in general chemistry textbooks.

Again, the presentation by Kotz and Purcell (1991) is very much in accord with the historical record, and the underlined part adds a new dimension with respect to the "debate raged in physics" and how scientists went about resolving this dilemma. On comparing different textbooks, a thoughtful student may wonder if these two textbooks (Segal 1989; Kotz and Purcell 1991) are presenting chemistry or history of chemistry. Indeed, this is the dilemma faced by most chemistry teachers and textbook authors. However, a critical appraisal of most of our current textbooks would show that if we want to understand chemistry, its history cannot be ignored. In other words, the history of chemistry is "inside" chemistry (Niaz and Rodriguez 2001).

All textbooks that were classified as Satisfactory (S) in this study clearly differentiated between the following two aspects: (a) origin of the wave–particle duality can be attributed to both Einstein and de Broglie based on existing problems at the beginning of the twentieth century (e.g., photoelectric effect and Bohr's model of the atom), and (b) experimental evidence based on diffraction of electron beams by crystals was found later. In contrast to the inductivist perspective, espoused by most science textbooks (and even teachers), this clearly presents to the students a novel way of conceptualizing progress in science. In other words, the dynamics of scientific progress imposes a particular sequence of events (history as part of science), and there is no one way of doing science (cf. Niaz 2009). Sometimes a theoretical idea needs to be substantiated through experimental evidence and alternatively the sequence of events may be inverted. Interestingly, after providing experimental details and diffraction patterns (Davisson and Germer), Silberberg (2000) concluded that this is an example of how a "… theoretical insight provides the impetus for an experimental test" (p. 275).

Teaching wave–particle duality as part of quantum theory is difficult as the latter itself is controversial due to a distinction between formalism and interpretation, which can facilitate understanding. According to Cushing (1991):

> The question is whether we are capable of truly *understanding* (or comprehending) quantum phenomena, as opposed to simply *accepting* the formalism and certain irreducible quantum correlations. The central issue is that of understanding versus merely redefining terms to paper over our ignorance. (p. 337, original italics)

In the educational context, some of these interpretations are Copenhagen interpretation, ensemble interpretation, and Bohmian mechanics (Cheong and Song 2014; Cushing 1994). Difficulties involved in teaching quantum theory at the introductory level are well documented in the literature (De Souza and Iyengar 2013; Garritz 2013; Greca and Freire 2014; Niaz and Fernández 2008; Tsaparlis and Papaphotis 2002, 2009). Interestingly, Padilla and Van Driel (2011) have suggested that the historical background of the wave–particle duality can be used to elucidate chemistry teacher's pedagogical content knowledge (Shulman 1986). The following is an example of a question they asked teachers of quantum chemistry at the undergraduate level in the Netherlands: "Could you tell how wave–particle duality was developed in the history of science? Do you pay attention to this historical development in your lessons" (p. 369). To go beyond the division between formalism and interpretation, Cheong and Song (2014) have suggested an agnostic (as suggested by one of the reviewers) framework based on suspension of judgment on the real behaviors of microscopic objects:

> … in which prediction rules and reality-related interpretations are distinguished. In the modified framework, the prediction rule category includes a set of equations and calculation rules for the prediction of phenomena. On the other hand, the reality- related interpretation is related to the claim of reality or the normative claim about the role of theory. This study considers wave function collapse as a calculation tool belonging to the prediction rule. (p. 1019)

Finally, textbook accounts can of course be enriched by including further details, such as: (i) Both Planck and Bohr are generally considered to be innovators (or even perhaps revolutionaries), and still they opposed Einstein and de Broglie with respect to the photoelectric effect and the wave–particle duality. (ii) Similarly, experimental work of Davisson, Germer, Thomson, Reid, and others that facilitated diffraction patterns was extremely difficult to interpret and required the participation of the scientific community (cf. Davisson and Germer 1927a; Thomson and Reid 1928). In other words some experiments (diffraction and interference) help us to understand one facet of nature and other experiments (photoelectric) yet another facet, and the two weave together to comprehend reality. On the contrary, students generally believe that experimental data unambiguously provide evidence for a particular theoretical framework without any controversy and conflict.

Chapter 8
Conclusions: From Empiricism to Historicism to Naturalism and Beyond

Introduction

The relationship between chemistry (science) education and the history and philosophy of science has a long history. Various scholars have blazed the trail of this endeavor, such as Conant (1947), Holton (1969), Klopfer (1969), and Robinson (1969). Gerald Holton's (1952) *Introduction to Concepts and Theories in Physical Science* provided a glimpse for students and teachers as to how science evolves through the interactions of theories, experiments, and the work of actual scientists within a history and philosophy of science (HPS) perspective. More recently, a new edition of this textbook has presented science as a human adventure, from Copernicus to Einstein and beyond (Holton and Brush 2001).

It seems that after this long association, the time is ripe for a critical appraisal and perhaps the need to look for alternatives. According to Duschl and Grandy (2013), over the last 100 years, there have been three major movements in philosophy of science: empiricism, historicism, and naturalized philosophy of science (especially the model-based cognitive view, see Chaps. 2 and 3 for details). Furthermore, they have suggested that as the historical turn is dated, science educators need to adopt the model-based view based on cognitive and social dynamics. One way of interpreting this suggestion is that historical reconstructions of science content are not as fruitful as many science educators would like us to believe (cf. Matthews 2015). At this juncture science educators are faced with the following scenario: from empiricism to historicism, we learned many things and it is time to move on and embrace naturalism. If we accept this, then there would be immediate consequences for the status and the extent to which historical aspects of science content would continue to play an important role in our classrooms. This is of necessity a dilemma for science educators.

M. Niaz, *Chemistry Education and Contributions from History and Philosophy of Science*, Science: Philosophy, History and Education,
DOI 10.1007/978-3-319-26248-2_8

Interestingly, a recent review of the literature in chemistry education has recognized the need for research in chemistry education based on history and philosophy of science:

> While the philosophy of chemistry is gradually emerging as a distinctive epistemology for chemistry … More work that investigates what constitutes the nature of chemistry—philosophically, epistemologically, and historically—and how it may become integrated into the curriculum is needed so that a better understanding of what chemistry education is all about may be obtained. (Teo et al. 2014, p. 20)

Application of the history and philosophy of science, and understanding the nature of science in the classroom, is however difficult. A review of the literature (Chap. 3) shows that there are four possible views for introducing the nature of science: consensus view (domain-general), model-based view (domain-specific), family resemblance view, and the integrated view. The family resemblance view is characterized by its lack of historical content, whereas the consensus view tries to highlight consensus aspects based on a review of the history and philosophy of science literature. Earman (2004) has argued that lack of consensus among philosophers of science with respect to basic issues (e.g., laws of nature) is so great that the situation is not only of "disagreement" but rather of "disarray" (see Chap. 2). In a similar vein, Hoyningen-Huene (2013) has concluded that there is no consensus among philosophers or historians or scientists about the nature of science (see Chap. 3). To make matters worse, two philosophers of science with close ties with the science education community have presented a somber picture: "Indeed, today philosophical views regarding various aspects and characteristics of science by relevant expert communities show a bewildering array of disparity and no sign of convergence. If anything, there is more divergence than ever before. There are raging disputes among realists, empiricists, constructivists, feminists, multiculturalists and postmodernists about the nature of science. Nor does it help to turn to scientists themselves who hold either rather naïve or else surprisingly diverse views in this regard" (Irzik and Nola 2011, p. 592).

Given this landscape in which not only it is difficult to follow historians and philosophers of science but also the understanding of scientists themselves about science is elusive, it is plausible to suggest an integration of various aspects of these disciplines. Precisely, in this context the integrated view of the nature of science suggests an integration of the domain-general and domain-specific aspects (see Chap. 3).

Going Beyond the Dilemma

Let us first have a close look at what a distinguished philosopher has said as to why he adopted naturalism: "How, as Kuhn suggested, could one use history of science to justify philosophy of science conclusions. I resolved this conflict for myself … [by] naturalizing the philosophy of science. This move puts the philosophy of science on the same naturalistic level as history of science" (Giere 2014; for the complete quote, see Chap. 2. This is the latest that I could find with respect to Giere's views on this subject). This clearly shows that Giere's conflict was that philosophy of science could not follow from history of science, and the two had to be placed at the same level. Nowhere does Giere state that historicism was dated, as suggested by Duschl and Grandy (2013). Actually, Giere (1999) has stated explicitly that the historical record helps to facilitate an appraisal of the reconstructions, of which Darwin's theory is the best exemplar (see Chap. 2 for details). Philip Kitcher (2007), another prominent philosopher, has endorsed the historical perspective in similar terms.

Similarly, Denis Phillips (2014) and Harvey Siegel (2014), two prominent philosophers of science with close ties with the science education community, consider that historical reconstructions can even potentially extend the naturalistic program (for details, see Chap. 2). Phillips points out further that like any other human enterprise, reconstructions in turn are also open to criticisms and revisions.

Interestingly, even Laudan (1996), a naturalist philosopher of science, would recommend the historical approach in categorical terms:

> The fact is that scientists do not need to study the history of their discipline to learn the Tradition; it is right there in every science textbook. It is not called history, of course. It is called 'science', but it is no less the historical canon for all that. Thus, the budding chemist learns Prout's and Avogadro's hypotheses, and Dalton's work on proportional combinations; he learns how to do Millikan's oil drop experiment; he works through Linus Pauling's struggles with the chemical bond … And *history's role in science pedagogy mirrors its centrality as gatekeeper of standards and methods*. (p. 153, italics added)

It seems that Laudan was writing the chemistry curriculum based on its history: Prout, Avogadro, Dalton, Millikan, and Pauling—all central figures in introductory chemistry courses (see Chaps. 4, 5, 6 and 7). The rationale for Laudan's approach is based on the fact that the sciences are much more tightly bound to their history than other intellectual activities (p. 153), so much so that the history of science is intertwined with the different topics of the science curriculum. Laudan's perspective provides further evidence for what we have referred to as: history of chemistry is "inside" chemistry (cf. Niaz and Rodríguez 2001; also see Chap. 1). However, analyses of science textbooks reveal that these are far from what Laudan would consider as "historical canon" (cf. Niaz 2014a for a recent review). Most textbooks simply reproduce biographical and anecdotal information with no reference to the conflicts and controversies in which the scientists had to defend their models and theories. Polanyi (1964) has highlighted the predicament by emphasizing the degree to which established knowledge in textbooks departs from the events associated with the original discovery:

> Yet as we pursue scientific discoveries through their consecutive publication on their way to the textbooks, which eventually assures their reception as part of established knowledge by successive generations of students, and through these by the general public, we observe that the intellectual passions aroused by them appear gradually toned down to a faint echo of their discoverer's first excitement at the moment of Illumination … A transition takes place here from a heuristic act to the routine teaching and learning of its results, and eventually to the mere holding of these as known and true, in the course of which the personal participation of the knower is altogether transformed. (pp. 171–172)

Given the present state of our textbooks in almost all parts of the world, and to follow Laudan's advice, research in chemistry (science) education based on a history and philosophy of science perspective is almost an imperative (cf. Niaz 2014a, b).

At its present juncture, science education does indeed face a dilemma (see Chap. 2). Science educators can either accept the advice of philosophers of science entirely, that is, naturalizing philosophy of science, or look for an alternative. One alternative is to follow historical reconstructions, case by case, based on the different topics of the curriculum, and draw tentative conclusions for teaching science. It seems that such an approach would be endorsed by some philosophers (e.g., Harvey Siegel and Denis Phillips). Even Giere would emphasize scientific practice based on perspectival rather than objective facets of science that requires looking back historically. Similarly, Laudan (1996) would endorse such an approach.

How to Integrate History of Chemistry with the Science Topic in the Classroom?

An integration of the history of chemistry with the science topic requires the understanding of (among other facets of the nature science) evidence, observations, inferences, arguments, and explanations. For example, it is necessary that students understand the distinction between data and evidence and are able to explain how data can be interpreted differently (i.e., the use of data as evidence) and how this is a potential source of bias. This is congruent with the NGSS (NRC 2013, Appendix F), which state:

> Being a critical consumer of information about science and engineering requires the ability to read or view reports of scientific or technological advances or applications (whether found in the press, the Internet, or in a town meeting) and to recognize the salient ideas, identify sources of error and methodological flaws, distinguish observations from inferences, arguments from explanations, and claims from evidence. (p. 13)

Again, *performance expectations* is one of the most innovative part of NGSS framework for improving science learning and can help teachers to design new teaching strategies:

> The real innovation in the NGSS is the requirement that students are required to operate at the intersection of practice, content, and connection. *Performance expectations* are the right way to integrate the three dimensions. It provides specificity for educators, but it also sets the tone for how science instruction should look in classrooms. If implemented properly, the NGSS will result in coherent, rigorous instruction that will result in students being able to acquire and apply scientific knowledge to unique situations as well as have the ability to think and reason scientifically. (NGSS, NRC 2013, Introduction, pp. 3–4)

Based on their CLUE curriculum, Cooper and Klymkowsky (2013) suggest including the following as a *performance expectation* (NRC 2013, NGSS) that combines science practice with disciplinary knowledge: "Using evidence from experiments, explain how and why models of atomic structure changed over time" (p. 1118). To implement this strategy would, of course, require an explicit historical reconstruction. In most parts of the world, general chemistry textbooks devote considerable space to the atomic models of Dalton, Thomson, Rutherford, Bohr, Bohr-Sommerfeld, and wave mechanical (see Chap. 4). As the experiments on which these atomic models are based are already described in textbooks, it is a sound advice to ask students to discuss and explore the possible reasons for which models have changed. This also clearly illustrates how the tentative nature of science (a domain-general aspect) can be directly integrated with domain-specific content knowledge.

The use of the alpha-particle scattering experiments in the early twentieth century was crucial in the postulation of the nuclear model of the atom. Most textbooks published in various parts of the world devote considerable space to these experiments and attribute the experiment to Rutherford and his colleagues in Manchester. However, almost all textbooks ignore that, on knowing Rutherford's results, Thomson did the same experiments at the Cavendish laboratory in Cambridge. Both groups of researchers obtained very similar results and still their interpretations were entirely different (see Chap. 3 and Niaz 2009 for details). In order to explain the results, Thomson propounded the hypothesis of compound scattering, and Rutherford propounded the hypothesis of single scattering. A bitter controversy ensued that lasted for many years. Given that our students are familiar with the historical details of the contributions of Thomson and Rutherford, inclusion of this controversy (as a *performance expectation*) can illustrate an important aspect of NOS (for teaching strategies based on these atomic models, see Niaz et al. 2002; Niaz 2011).

The periodic table of chemical elements constitutes an important and essential part of all introductory courses and general chemistry textbooks. Given the long and interesting development of its history, the following aspects can easily be adapted for classroom discussions (cf. Niaz and Luiggi 2014): (a) Is it possible that for almost 100 years (1820–1920), scientists had no idea or never asked the question as to whether there could be an underlying rationale for explaining periodicity? (b) Which is more important accommodation or prediction in the classification of the elements? (c) Which of the following explains better the role of periodicity in the origin of the periodic table: (i) inductive generalization or (ii) atomic theory? (d) Mendeleev's contribution can be best considered as a (i) classification scheme, (ii) empirical law, or (iii) theoretical framework. (e) The periodic table has a long history and was developed by (i) Mendeleev or (ii) various scientists that formed part of the scientific community. Discussion of the pros or cons of these statements can help students to generate arguments in support of their particular position and understand that in order to achieve consensus scientists go through a complex process of scrutiny of knowledge claims.

Electrolyte solution chemistry forms part of most introductory general chemistry textbooks. A historical reconstruction of the topic shows that during the late nineteenth and early twentieth century, there was considerable controversy among the ionist and hydrationist schools of chemists in order to explain the phenomenon of osmosis (see Chap. 7 for details). According to De Berg (2014b), this controversy can form the background for a debate in the classroom in order to illustrate how the same data can be explained by different theoretical frameworks (an important domain-general NOS aspect).

Cooper et al. (2010) found that providing students with "foolproof" rules for writing the Lewis structures is not very helpful. Generally, in most introductory courses, students first learn the ionic bond that involves transfer of electrons. On the other hand, Lewis structures were a device to understand covalent bonds based on sharing of electrons that is counterintuitive for students. Even Lewis's colleagues and chemists at first found considerable difficulties with respect to sharing of electrons in a covalent bond (see Chaps. 6 and 7 and Niaz 2001c for details). A major argument against Lewis's idea of sharing electrons was that the approach of two electrons having the same charge should produce repulsive forces and hence produce destabilization. An explanation of how this difficulty was overcome by Lewis and others can help arouse students' interest and curiosity and thus lead to a better understanding of Lewis structures. In other words, rules and algorithms will make sense to the students only if these are presented in the context of the development of the topic. In many cases this context becomes explicit during a historical reconstruction of the events that led to its development.

In understanding the photoelectric effect, students manifest a major alternative conception: the longer the time taken by the incident light, the greater the frequency and the more energy is stored in electrons inside the metal, leading to higher kinetic energy of the ejected electrons (see Oh 2011 and Chap. 7 for details). This is quite similar to the trigger hypothesis suggested by Lenard (1902) to explain the photoelectric effect. A classroom teaching strategy that juxtaposes students' alternative conceptions with Lenard's in the context of Einstein's explanation of the photoelectric effect can help to produce cognitive conflict and facilitate greater conceptual understanding.

The development of wave–particle duality in the history of science can help illustrate various aspects of quantum mechanics (see Chap. 7). Furthermore, classroom discussions around the following historical aspects can help to facilitate understanding of: (a) Why was it difficult to design experiments for measuring matter waves? (b) Why did the scientific community oppose de Broglie's ideas and still accepted the same ideas presented by Schrödinger? (c) Why did Planck and Bohr oppose Einstein and de Broglie in the development and acceptance of not only wave–particle duality but also the photoelectric effect?

The scientific method as a part of the domain-general nature of science is perhaps one of the most difficult aspect for students, teachers, and even perhaps curriculum developers. Consequently, its integration within science content is even more important and may even help to facilitate a better understanding. Here I would like to summarize results from five different studies (that explore the scientific method) conducted in different countries and reported in different parts of this book:

(a) Lederman et al. (2002) suggested that the scientific method was a myth.
(b) Osborne et al. (2003), based on the expert community in the UK, suggested that the scientific method represents the central thrust of scientific research.
(c) Windschitl (2004), based on research on preservice secondary science teachers in the USA, suggested considerable difficulties in understanding the scientific method. Interestingly, the author suggested that perhaps the science education

community itself promoted it (albeit subtly, as some science educators still consider it to be necessary).

(d) Wan et al. (2013) based on science teacher educators from mainland China also found difficulties, although some participants had an informed view of the scientific method.

(e) Niaz (2016, this book, Chap. 3) based on in-service science teachers in Venezuela also found difficulties. However, some students had an informed view, and here is an example: "The freedom of a scientist cannot be curtailed to a degree that makes him follow only one method—there must always be some indication of an idea, a hypothesis—leading to multiple ways of finding expected or unforeseen results that may generate great discoveries" (Participant #10, study reported in Chap. 3). Another participant even suggested a means of integrating the scientific method within science content: "There is evidence that Millikan discarded data obtained in his experiment, which means that he did not follow or respect the scientific method rigorously and still his findings are to this day accepted by the scientific community" (Participant #2, study reported in Chap. 3).

These studies provide evidence of the difficulties involved in understanding the scientific method and how the topic may be explored in future research. Furthermore, following the advice of NGSS, it is important to note that the performance expectations (e.g., scientific method) will have to be implemented by instructional strategies designed by the teachers themselves:

> The NGSS are standards, or goals, that reflect what a student should know and be able to do—they do not dictate the manner or methods by which the standards are taught. The performance expectations are written in a way that expresses the concept and skills to be performed but still leaves curricular and instructional decisions to states, districts, school and teachers. The performance expectations do not dictate curriculum; rather, they are coherently developed to allow flexibility in the instruction of the standards. (NGSS, NRC 2013, Executive summary, p. 2)

One example of such a teaching strategy is the recent suggestion by Binns and Bell (2015) that the reference to the scientific method itself be avoided and that teachers could instead refer to the work of scientists as inquiry and explicitly describe the investigation conducted by the scientists.

Methodological Pluralism

C.A. Coulson was a major figure in the development of the molecular orbital theory to explain chemical bonding during the first half of the twentieth century (see Chap. 6 for details). Coulson was originally trained in mathematics and physics, and still he was sensitive in his writings with respect to the chemist's needs for visualization of chemical bonds based on L. Pauling's concept of resonance. Although Coulson made no secret of his preference for the molecular orbital theory, he considered both theories (molecular orbital and valence bond) to be approximations and recommended both. According to Gavroglu and Simões (2012):

> More than anyone else he [Coulson] was a stubborn and committed advocate of *methodological pluralism*, of the possibilities for exploring different approaches in different problems, always eager to compare and contrast them, to foster semiempirical calculations while at the same time exploring the potential of ever more potent computers, all within the overarching view that privileged conceptual understanding over numerical accuracy. (pp. 226–227, emphasis added)

It is because of these beginnings and efforts that at present, after more than 70 years, both valence bond and molecular orbital theories continue to form part of the intellectual heritage of chemistry (Hoffmann et al. 2003).

With this background, let us consider Giere's (2006a, b, 2014) naturalism (see Chap. 2 for details). Giere (2010, p. 214) characterizes naturalism not as a thesis but as a method. Furthermore, knowledge claims are perspectival rather than absolutely objective and hence cannot provide a "true" or "correct" answer to a problem. Consequently, instead of postulating one method or theory for a problem, Giere (2006b) has recommended a *pluralism of perspectives*, at different levels. It is this *pluralism of perspectives* that led to *methodological pluralism*, facilitating a better understanding of chemical bond formation based on two different and rival perspectives, namely, valence bond and molecular orbital. It is suggested that methodological pluralism is an important guide for understanding scientific progress in the history of science. Actually, Giere (2010) was quite explicit in recommending that: "What, one might reasonably ask, constitutes a scientific account of anything? The best general answer a naturalist can give is: *A scientific account is one sanctioned by a currently recognized science*" (p. 212, italics in the original). The 'currently recognized science' would of course refer to the scientific community, engaged in 'science in the making' (cf. Niaz 2012a).

Let us now go back and ask the following question for chemical bond formation: Which account was sanctioned in the 1950s and at present? Both valence bond and molecular orbital (cf. Hoffmann et al., 2003; Shaik and Hiberty 2008; also Chap. 6). Now let us extend this argument a little further: Which account for understanding nature of science in science education is sanctioned by the scientific community? My tentative response, which is open to objections, is all four: domain-general (consensus based), domain-specific (model based), family resemblance, and integrated views.

True Theories, Pessimistic Induction, Contingency, and Tentative Nature of Science

Apparently, teaching the tentative nature of scientific knowledge should be the least controversial part of teaching the nature of science and chemistry or for that matter any science subject. This precisely provides the opportunity to share with the students what history of science shows us, namely, that all theories will eventually change (in other words, there are no "true" theories). However, it seems that some science educators consider historicism to be dated and thus avoid the pessimistic induction, leading them to state that if scientific inquiry does not lead to truths, this may cause students to lose confidence in science (Duschl and Grandy 2013). This can be countered on the grounds, that if theories change, then students can be encouraged to contemplate the following scenario: Despite all the progress in science based on the hard work, perseverance, creativity, and imagination of many scientists (not necessarily geniuses), a lot remains to be done and our present-day students can contribute toward the next step (theory) in scientific progress.

I wonder how would a chemistry teacher respond to a student who might ask the following question after studying the chapter on atomic models: Why did the atomic models of Thomson, Rutherford, Bohr, and Bohr-Sommerfeld change in quick succession in the early twentieth century? Another student might even go further and ask: If one atomic model is replaced, does it mean that the new model is true and the previous was wrong? These questions can come up in almost any introductory level course in any part of the world and do not necessarily form part of the history of science. According to Laudan (1996), this constitutes "science" and not necessarily the "historical canon." In other words, these questions will have to be answered by scientists and science educators and not just the historians.

Let us now seek help from another naturalist philosopher of science with respect to our dilemma:

> Nineteenth-century Newtonian physicists were surely as justified in thinking that they had discovered the objectively real structure of the world as any scientist could possibly be. Yet Newtonian gravitational theory has been abandoned as the accepted account of gravitational phenomena (though not for many applications for which it yields sufficiently accurate predictions). Can it be anything more than *presentist hubris* to think that we now have the objectively correct theory? (Giere 2006a, p. 95, italics added)

I am sure Giere's response would have been the same in the case of atomic models and for that matter for so many other topics of the science curriculum.

A related question was addressed in Chap. 5 in the context of learning stoichiometry: Do scientific laws help in learning science? It was concluded that it was the context of solving a problem (with appropriate guidelines) that can help students to understand the underlying stoichiometric relationship. On the contrary, most textbook authors and teachers simply emphasize the memorization of the laws and then ask students to solve algorithmic problems. The same teaching strategy is followed for almost all the topics of the curriculum. For example, in the case of chemical bonding, Cooper et al. (2010) have shown that the rules for writing Lewis structures do not facilitate conceptual understanding. In effect, these rules in general are not presented in the *context* of the development of a topic. Most philosophers and historians of science do recognize the role played by laws and for that matter also theories. However, in the educational context, overemphasizing the role of laws and theories can be counterproductive. According to Giere (2006a), to question the applicability of the laws of nature is not to question any science itself but rather go beyond and understand the aims and achievements of scientific activities (pp. 69–70).

The contingency thesis (cf. Chap. 6) can provide grounds for a possible competition among rival theories and is yet another aspect of the nature of science that can provide opportunities for discussion in the classroom. How can we explain to students that both valence bond and molecular orbital theories can successfully explain various aspects of the covalent bond and hence both theories are "correct" (Gavroglu and Simões 2012)? In a survey based on practicing chemists, Brush (1999) found that both valence bond and molecular orbital theories provide almost equally good descriptions of benzene and similar molecules. Recapitulating the contingency thesis, according to Cushing (1994): "Even in situations in which there may be observationally equivalent theories (such as the Copenhagen and Bohm versions of quantum mechanics), who gets to the top of the hill first holds the high ground and must be dislodged (if required, not otherwise)" (p. 5). Interestingly, in the case of valence bond and molecular orbital theories, both got to the top of the hill at about the same time and have been there for almost 70 years, and hence the chemical community has been saved from deciding whom to "dislodge."

At this stage, it would be interesting to discuss the following issue raised by one of the reviewers of this book: "The competition among rival theories does not imply the contingency of scientific development. To demonstrate that one would have to show that theories get accepted for contingent (i.e., non-epistemic) reasons." In a sense, the reviewer is correct. However, we need to discuss the two events (covalent bonding and quantum mechanics) discussed in this book to illustrate the contingency thesis within a wider historical perspective. In the case of covalent bonding, we have ample evidence to show that what started as a contingency finally led to an unending rivalry between the two interpretations (for details, see Chap. 6 and Hoffmann et al. 2003). However, the case of quantum mechanics is more complex and requires a historical background. According to Bell (1987), a leading scholar on the Bohmian interpretation of quantum mechanics:

> But in 1952 I saw the impossible done. It was in papers by David Bohm. Bohm showed explicitly how parameters could indeed be introduced, into nonrelativistic wave mechanics, with the help of which the indeterministic description could be transformed into a deterministic one. More importantly, in my opinion, the subjectivity of the orthodox version, the necessary reference to the 'observer,' could be eliminated. … But why then had Born not told me of this 'pilot wave'? If only to point out what was wrong with it? Why did von Neumann not consider it? More extraordinarily, why did people go on producing "impossibility" proofs, after 1952, and as recently as 1978? … Why is the pilot wave picture ignored in text books? Should it not be taught, not as the only way, but as an antidote to the prevailing complacency? To show us that vagueness, subjectivity, and indeterminism, are not forced on us by experimental facts, but by deliberate theoretical choice? (p. 160)

Max Born and von Neumann were two leading quantum physicists who critiqued Bohm's interpretation and thus helped to establish the hegemony of the Copenhagen interpretation of quantum mechanics. Bell asks two very pertinent questions: (a) Why is the pilot wave picture (de Broglie and Bohm) ignored in textbooks? (b) Should Bohm's interpretation not be taught? These are difficult questions, and to the best of my knowledge, the late James Cushing did not consider them. The closest that Cushing (1996) came to was when he criticized the textbook by Landau and Lifshitz (1958) for stating that, "… In quantum mechanics there is no such concept as the path of a particle [Bohmian mechanics]" (p. 2).

More recently, Lautesse et al. (2015) have referred to a possible rivalry in quantum mechanics between the conservative position (Copenhagen school) and the innovative position based on "quantons." The latter is based on the work of Bunge (2003) and Lévy-Leblond (1988). Similarly, Laloë (2001), a physicist, has expressed views (based on an extensive review of the literature, see Chap. 6) that can be considered as a possible rivalry between the Copenhagen and Bohmian mechanics based on "hidden variables" (Bohm 1952). According to Sheldon (2013), it is the positions of the particles in Bohmian mechanics that are its "hidden variables." In other words, in quantum mechanics, the scientific community subscribes to three rival interpretations: Copenhagen, Bohmian, and "quantons." To conclude, competition among rival theories does not imply the contingency thesis, but the thesis itself can provide grounds for rivalry among the three competing theories.

After going through empiricism, historicism, and naturalism, perhaps it is time to look at the horizon for alternatives. In this quest, the following advice from Cushing (1994) can be thought-provoking: "… as a pragmatic matter, we can simply choose, from among the consistent, empirically adequate theories on offer at any time, that one which allows us best to 'understand' the phenomena of nature, while not confusing this practical virtue with any argument for the 'truth' or faithfulness of representation of the story thus chosen" (p. 215). It is plausible to suggest that one alternative would be to follow the naturalist stance along with in-depth historical reconstructions, based on Giere's *pluralism of perspectives* and Coulson's *methodological pluralism*. This *plurality of models* has also been endorsed by Kellert et al. (2006): "Seeking a proper plurality of models, each of which accurately accounts for some but not all aspects of the situation, might be preferable. What is the advantage of the pluralist interpretation? … it provides a means of avoiding senseless controversies that do not lead to progress. It also helps emphasize the partiality of scientific knowledge" (p. xv).

Appendices

Appendix 1: List of General Chemistry Textbooks Published in the USA, Analyzed in Different Chapters of This Book (n = 72)

1. Ander, P., & Sonnessa, A. (1968). *Principles of chemistry* (Spanish ed.). New York: Macmillan.
2. Armstrong, J. (2012). *General, organic, and biochemistry: An applied approach.* Belmont: Brooks/Cole (Cengage).
3. Atkins, P., & Jones, L. (2002). *Chemical principles: The quest for insight* (2nd Ed.). New York: Freeman.
4. Atkins, P., & Jones, L. (2008). *Chemical principles: The quest for insight* (4th Ed.). New York: Freeman.
5. Bettelheim, F.A., Brown, W.H., Campbell, M.K., Farrell, S.O., & Torres, O.J. (2012). *Introduction to general, organic and biochemistry* (9th Ed.). Belmont: Brooks/Cole (Cengage).
6. Bishop, M. (2002). *An introduction to chemistry.* San Francisco: Benjamin Cummings.
7. Bodner, G., & Pardue, H. (1989). *Chemistry: An experimental science.* New York: Wiley.
8. Brady, J., & Humiston, G. (1996). *General chemistry: principles and structure* (Spanish ed.). New York: Wiley.
9. Brady, J., Russell, J., & Holum, J. (2000). *Chemistry: The study and its changes* (3rd Ed.). New York: Wiley.
10. Brown, L., & Holme, T. (2011). *Chemistry for engineering students* (2nd Ed.). Belmont: Brooks/Cole (Cengage).
11. Brown, T.L., Le May, H.E., & Bursten, B. (1997). *Chemistry: The central science* (7th Ed., Spanish). Englewood Cliffs: Prentice Hall.

M. Niaz, *Chemistry Education and Contributions from History and Philosophy of Science*, Science: Philosophy, History and Education,
DOI 10.1007/978-3-319-26248-2

12. Brown, T.L., LeMay, H.E., Bursten, B.E., & Murphy, C.J. (2012). *Chemistry: The central science* (12th Ed.). Englewood Cliffs: Prentice Hall (Pearson Education).
13. Burns, R. (1996). *Fundamentals of chemistry* (Spanish ed.). Englewood Cliffs: Prentice Hall.
14. Chang, R. (1998). *Chemistry* (6th Ed., Spanish). New York: McGraw-Hill.
15. Chang, R. (2010). *Chemistry* (10th Ed., Spanish). New York: McGraw-Hill.
16. Cracolice, M.S., & Peters, E.I. (2012). *Introductory chemistry* (5th Ed.). Belmont: Brooks/Cole (Cengage).
17. Daub, G.W., & Seese, W. (1996). *Basic chemistry* (8th Ed., Spanish). Englewood Cliffs: Prentice Hall.
18. Denniston, K.J., Topping, J.J., & Caret, R.L. (2011). *General, organic, and biochemistry* (7th Ed.). New York: McGraw-Hill.
19. Dickerson, R.E., Gray, H.B., Darensbourg, M., & Darensbourg, D. (1970). *Chemical principles* (4th Ed.). Menlo Park: Benjamin Cummings.
20. Dickerson, R.E., Gray, H.B., & Haight, G.P. (1979). *Chemical principles.* Menlo Park: Benjamin Cummings.
21. Dickson, T. (2000). *Introduction to chemistry* (8th Ed.). New York: Wiley.
22. Ebbing, D.D. (1996). *General chemistry* (5th Ed., Spanish). New York: McGraw-Hill.
23. Ebbing, D.D., & Gammon, S.D. (2012). *Chemistry* (10th Ed.). Belmont: Brooks/Cole (Cengage).
24. Fine, L.W., & Beall, H. (1990). *Chemistry for engineers and scientists.* Philadelphia: Saunders.
25. Frost, L., Deal, T., & Timberlake, K.C. (2011). *General, organic and biological chemistry: An integrated approach.* Upper Saddle River: Pearson Prentice Hall.
26. Goldberg, D. (2001). *Fundamentals of chemistry* (3rd Ed.). New York: McGraw-Hill.
27. Hein, M. (1990). *Foundations of college chemistry* (Spanish ed.). Pacific Grove: Brooks/Cole).
28. Hein, M., & Arena, S. (1997). *Foundations of college chemistry.* Belmont: Brooks/Cole.
29. Hill, J. (1975). *Chemistry for changing times* (2nd Ed.). Minneapolis: Burgess Publishing Company.
30. Hill, J., & Petrucci, R. (1999). *General chemistry: An integrated approach* (2nd Ed.). Upper Saddle River: Prentice Hall.
31. Holtzclaw, H.F., & Robinson, W.R. (1988). *General chemistry* (8th Ed.). Lexington: Heath.
32. Jones, L., & Atkins, P. (2000). *Chemistry: Molecules, matter and change* (4th Ed.). New York: Freeman.
33. Kotz, J.C., Treichel, P.M., & Townsend, J. (2010). *Chemistry and chemical reactivity* (7th Ed.). Belmont: Brooks/Cole (Cengage).
34. Lippincott, W.T., Garrett, A., & Verhoek, F. (1977). *Chemistry: A study of matter* (3rd Ed.). New York: Wiley.

35. Mahan, B., & Myers, R. (1990). *University chemistry* (4th Ed., Spanish). Menlo Park: Benjamin Cummings.
36. Malone, L. (2001). *Basic concepts of chemistry* (6th Ed.). New York: Wiley.
37. Masterton, W.L., & Hurley, C.N. (2009). *Chemistry: Principles and reactions* (6th Ed.). Belmont: Brooks/Cole (Cengage).
38. Masterton, W.L., & Slowinski, E.J. (1977). *Chemical principles* (4th Ed.). Philadelphia: Saunders.
39. McMurry, J., Castellion, M.E., & Ballantine, D.S. (2007). *Fundamentals of general, organic and biological chemistry* (5th Ed.). Upper Saddle River: Pearson Prentice Hall.
40. McMurry, J., & Fay, R. (2001). *Chemistry* (3rd Ed.). Upper Saddle River: Prentice Hall.
41. McMurry, J., & Fay, R. (2012). *Chemistry* (6th Ed.). Upper Saddle River: Prentice Hall.
42. McQuarrie, D.A., Rock, P.A., & Gallogly, E.B. (2011). *General chemistry* (4th Ed.). Mill Valley: University Science Books.
43. Miller, F.M. (1984). *Chemistry: Structure and dynamics.* New York: McGraw-Hill.
44. Moore, J.W., Stanitski, C.L., & Jurs, P.C. (2002). *Chemistry: The molecular science.* Orlando: Harcourt College Publishers.
45. Mortimer, C. (1983). *Chemistry* (5th Ed.). Belmont: Wadsworth.
46. O'Connor, R. (1972). *Fundamentals of chemistry: A learning systems approach.* New York: Harper & Row.
47. Oxtoby, D., Gillis, H.P., & Nachtrieb, N. (1999). *Principles of modern chemistry* (4th Ed.). Philadelphia: Saunders.
48. Oxtoby, D., Gillis, H.P., & Campion, A. (2012). *Principles of modern chemistry* (7th Ed.). Belmont: Brooks/Cole (Cengage).
49. Oxtoby, D., Nachtrieb, N., & Freeman, W. (1990). *Chemistry: Science of change.* (2nd Ed.). Philadelphia: Saunders.
50. Parry, R.W., Steiner, L.E., Tellefsen, R.L., & Dietz, P.M. (1970). *Chemistry: Experimental foundations*. New Jersey: Prentice-Hall.
51. Petrucci, R.H. (1972). *General chemistry: Principles and modern applications.* New York: Macmillan.
52. Quagliano, J.V., & Vallarino, L.M. (1969). *Chemistry* (3rd Ed.). Englewood Cliffs: Prentice Hall.
53. Raymond, K.W. (2010). *General, organic and biological chemistry: An integrated approach* (3rd Ed.). New York: Wiley.
54. Russo, S., & Silver, M. (2002). *Introductory chemistry* (2nd Ed.). San Francisco: Benjamin Cummings.
55. Seager, S.L., & Slabaugh, M.R. (2013). *Chemistry for today: General, organic and biochemistry* (7th Ed.). Belmont: Brooks/Cole (Cengage).
56. Segal, B. (1989). *Chemistry: Experiment and theory* (2nd Ed.). New York: Wiley.
57. Silberberg, M. (2000). *Chemistry: The molecular nature of matter and change* (2nd Ed.). New York: McGraw-Hill.

58. Sisler, H., Dresdner, R., & Mooney, W. (1980). *Chemistry: A systematic approach.* New York: Oxford University Press.
59. Spencer, J.N., Bodner, G.M., & Rickard, L.H. (2012). *Chemistry: Structure and dynamics* (5th Ed.). New York: Wiley.
60. Stoker, H.S. (1990). *Introduction to chemical principles* (3rd Ed.). New York: McMillan.
61. Stoker, H.S. (2010). *General, organic, and biological chemistry* (5th Ed.). Belmont: Brooks/Cole (Cengage).
62. Timberlake, K.C. (2010). *General, organic and biological chemistry* (3rd Ed.). Upper Saddle River: Pearson Prentice Hall.
63. Tro, N. (2008). *Chemistry: A molecular approach.* Upper Saddle River: Prentice Hall (Pearson).
64. Tro, N. (2012). *Chemistry in focus: A molecular view of our world.* Belmont: Brooks/ Cole (Cengage).
65. Umland, J., & Bellama, J. (1999). *General chemistry* (3rd Ed.). Pacific Grove: Brooks/Cole.
66. Whitten, K.W., Davis, R.E., Peck, M.L., & Stanley, G.G. (2013). *Chemistry* (10th Ed.). Belmont: Brooks/Cole (Cengage).
67. Whitten, K.W., Gailey, K.D., & Davis, R.E. (1992). *General chemistry* (3rd Ed., Spanish). Philadelphia: Saunders.
68. Wolfe, D. (1988). *Introduction to college chemistry* (2nd Ed.). New York: McGraw-Hill.
69. Zumdahl, S.S. (1989). *Chemistry* (2nd Ed.). Lexington, MA: Heath.
70. Zumdahl, S.S. (1990). *Introductory chemistry: A foundation.* Lexington: Heath.
71. Zumdahl, S.S., & Zumdahl, S.A. (2012). *Chemistry: An atoms first approach.* Belmont: Brooks/Cole (Cengage).
72. Zumdahl, S.S., & Zumdahl, S.A. (2014). *Chemistry* (9th Ed.). Belmont: Brooks/ Cole (Cengage).

Appendix 2: List of General Chemistry Textbooks Published in Turkey, Analyzed in This Book (n = 27)

1. Alpaydın, S., & Şimşek, A. (2006). *Genel kimya* (2. Baskı). Ankara: Nobel Yayın Dağıtım.
2. Aydın, A.O., Sevinç, V., & Şengil, İ.A. (2001). *Temel kimya* (2. Baskı). Adapazarı: Aşiyan Yayınları.
3. Atasoy, B. (2000). *Genel kimya.* Ankara: Gündüz Eğitim ve Yayıncılık.
4. Atasoy, B. (2004). *Temel kimya kavramları* (2. Baskı). Ankara: Asil Yayın Dağıtım.
5. Bağ, H. (2006). *Genel kimya-I* (1. Baskı). Ankara: Pegem A Yayıncılık.
6. Bayın, Ö. (1982). *Modern kavramlar yaklaşımıyla kimya.* İstanbul: Fil Yayınevi.

7. Baykut, F. (1964). *Modern denel, genel kimya dersleri.* İstanbul: İstanbul Üniversitesi Yayınları.
8. Bekaroğlu, Ö., & Tan, N. (1986). *Genel kimya (teori ve problemler).* İstanbul: Kipaş Dağıtımcılık.
9. Dikman, E. (1975). *Temel kimya (anorganik).* İzmir: Ege Üniversitesi Fen Fakültesi Yayınları.
10. Erdik, E., & Sarıkaya, Y. (1991). *Temel üniversite kimyası* (5. Baskı). Ankara: Hacettepe-Taş Kitapçılık Ltd. Şt.
11. Ergül, S. (2006). *Genel kimya.* Ankara: Anı Yayıncılık.
12. Hakdiyen, İ. (1960). *Genel ve teknik kimya.* İstanbul: Teknik Okulu Yayınları.
13. Hazer, B. (1997). *Genel kimya.* Trabzon: Akademi Ltd. Şti.
14. İrez, G. (2002). *Temel kimya-1.* Muğla: Muğla Üniversitesi Yayınları.
15. Öncel, M.F. (1974). *Genel kimya notları-1,* Ankara.
16. Öncel, M. F. (1976). *Deney ve problemleri ile modern genel kimya-1.* Ankara: Fen Yayınevi.
17. Özcan, M. (1998). *Modern temel kimya-I* (Genişletilmiş 2. Baskı). Balıkesir: Vipaş Yayınları.
18. Pamuk, F. (1984). *Genel kimya.* Ankara: Gazi Üniversitesi Yayınları.
19. Saracoğlu, A.S. (1983). *Temel kimya* (3. Baskı). İstanbul: Çağlayan Kitabevi.
20. Saraç, A.S., Güvençoğlu, A., & Soydan, A.B. (1983). *Modern genel kimya ve çözümlü problemleri.* İstanbul: Murat Matbaacılık.
21. Soydan, B., & Saraç, A.S. (1998). *Genel üniversite kimyası ve modern uygulamaları* (2. Baskı). İstanbul: Seç Yayın Dağıtım.
22. Şenvar, C. (1989). *Temel kimya.* Ankara: Hacettepe Üniversitesi Yayınları.
23. Tosun, F. (1969). *Genel kimya, prensipler.* Trabzon: Karadeniz Teknik Üniversitesi Yayınları.
24. Tunalı, N.K., & Aras, N.K. (1977). *Kimya temel kavramlar* (11. Baskı). Ankara: Başarı Yayınları.
25. Ün, R. (1967). *Genel kimya (Genel ve anorganik).* Trabzon: Karadeniz Teknik Üniversitesi Yayınları.
26. Ünal, S. (1992). *Genel kimya.* İstanbul: Marmara Üniversitesi Yayınları.
27. Yavuz, O. (1978). *Genel kimya.* Erzurum: Atatürk Üniversitesi Basımevi.

References

Abd-El-Khalick, F. (2005). Developing deeper understandings of nature of science: The impact of a philosophy of science course on preservice science teachers' views and instructional planning. *International Journal of Science Education, 27*, 15–42.
Abd-El-Khalick, F. (2012). Examining the sources for our understandings about science: Enduring conflations and critical issues in research on nature of science in science education. *International Journal of Science Education, 34*(3), 353–374.
Abd-El-Khalick, F., Waters, M., & Le, A. (2008). Representation of nature of science in high school chemistry textbooks over the past four decades. *Journal of Research in Science Teaching, 45*, 835–855.
Achieve, NGSS. (2013). Next generation science standards. Retrieved July 25, 2014 from http://www.nextgenscience.org/nextgeneration-science-standards
Achinstein, P. (1987). Scientific discovery and Maxwell's kinetic theory. *Philosophy of Science, 54*, 409–434.
Achinstein, P. (1991). *Particles and waves: Historical essays in the philosophy of science*. New York: Oxford University Press.
Adey, P., & Shayer, M. (1994). *Really raising standards: Cognitive intervention and academic achievement*. London: Routledge.
Adey, P., Shayer, M., & Yates, C. (2001). *Thinking science: The curriculum materials of the CASE project* (3rd ed.). London: Routledge.
Agung, S., & Schwartz, M. S. (2007). Students' understanding of conservation of matter, stoichiometry and balancing equations in Indonesia. *International Journal of Science Education, 29*(3), 1679–1702.
Akerson, V. L., Abd-El-Khalick, F., & Lederman, N. G. (2000). Influence of a reflective explicit activity-based approach on elementary teachers' conceptions of nature of science. *Journal of Research in Science Teaching, 37*, 295–317.
Alters, B. J. (1997). Whose nature of science? *Journal of Research in Science Teaching, 34*, 39–55.
American Association for the Advancement of Science, AAAS. (1993). *Benchmarks for science literacy: Project 2061*. Washington, DC: Oxford University Press.
Arabatzis, T. (2006). *Representing electrons: A biographical approach to theoretical entities*. Chicago: University of Chicago Press.
Armstrong, H. E. (1927). Poor common salt. *Nature, 120*, 478.
Armstrong, H. E. (1928). The nature of solutions. *Nature, 121*(3037), 48–51.
Arriassecq, I., & Greca, I. M. (2007). Approaches to the teaching of special relativity theory in high school and university textbooks of Argentina. *Science & Education, 16*, 65–86.

M. Niaz, *Chemistry Education and Contributions from History and Philosophy of Science*, Science: Philosophy, History and Education,
DOI 10.1007/978-3-319-26248-2

Atkins, P., & Jones, L. (2002). *Chemical principles: The quest for insight* (2nd ed.). New York: Freeman.
Atkins, P., & Jones, L. (2008). *Chemical principles: The quest for insight* (4th ed.). New York: Freeman.
Bektas, O., Ekiz, B., Tuysuz, M., Kutucu, E. S., Tarkin, A., & Uzuntiryaki-Kondakcib, E. (2013). Pre-service chemistry teachers pedagogical content knowledge of the nature of science in the particulate nature of matter. *Chemistry Education Research and Practice, 14*, 201–213.
Bell, J. S. (1987). *Speakable and unspeakable in quantum mechanics*. Cambridge, UK: Cambridge University Press.
Bensaude-Vincent, B. (1986). Mendeleev's periodic system of chemical elements. *British Journal for the History of Science, 19*, 3–17.
Bensaude-Vincent, B. (2014). Philosophy *of* chemistry or philosophy *with* chemistry? *Hyle, 20*(1), 58–76.
Bevilacqua, F., & Bordoni, S. (1998). New contents for new media: Pavia project physics. *Science & Education, 7*, 451–469.
Bidell, T. (1988). Piaget, Vygotsky and the dialectic of development. *Human Development, 31*, 329–348.
Binns, I. C., & Bell, R. L. (2015). Representation of scientific methodology in secondary science textbooks. *Science & Education, 24*, 913–936.
Blanco, R., & Niaz, M. (1997). Epistemological beliefs of students and teachers about the nature of science: From 'Baconian inductive ascent' to the 'irrelevance' of scientific laws. *Instructional Science, 25*, 203–231.
Blanco, R., & Niaz, M. (1998). Baroque tower on a gothic base: A Lakatosian reconstruction of students' and teachers' understanding of structure of the atom. *Science & Education, 7*, 327–360.
Blanco, E., & Niaz, M. (2014). Venezuelan university students' understanding of the nature of science. *Journal of Science Education, 15*(2), 66–70.
Bohm, D. (1952). A suggested interpretation of the quantum theory in terms of 'hidden' variables, I. *Physical Review, 85*, 166–179.
Bohm, D. (1980). *Wholeness and the implicate order*. London: Routledge & Kegan Paul.
Bohr, N. (1913). On the constitution of atoms and molecules. Part 1. *Philosophical Magazine, 26*(Series 6), 1–25.
BouJaoude, S., & Barakat, H. (2003). Students' problem solving strategies in stoichiometry and their relationship to conceptual understanding and learning approaches. *Electronic Journal of Science Education, 7*(3), 1–42.
BouJaoude, S., Salloum, S., & Abd-El-Khalick, F. (2004). Relationships between selective cognitive variables and students' ability to solve chemistry problems. *International Journal of Science Education, 26*(1), 63–84.
Brito, A., Rodríguez, M. A., & Niaz, M. (2005). A reconstruction of development of the periodic table based on history and philosophy of science and its limitations for general chemistry textbooks. *Journal of Research in Science Teaching, 42*, 84–111.
Brock, W. H. (1993). *The Norton history of chemistry*. New York: W.W. Norton.
Brown, J. R. (1990). Proof and truth in Lakatos's masterpiece. *International Studies in the Philosophy of Science, 4*(2), 117–130.
Bruner, J. (1977). *The process of education*. Cambridge, MA: Harvard University Press.
Brush, S. G. (1974). Should the history of science be rated X. *Science, 183*, 1164–1172.
Brush, S. G. (1976). *The kind of motion we call heat: A history of the kinetic theory of gases in the 19th century*. New York: North-Holland.
Brush, S. G. (1978). Why chemistry needs history and how it can get some. *Journal of College Science Teaching, 7*, 288–291.
Brush, S. G. (1989). History of science and science education. *Interchange, 20*, 60–70.
Brush, S. G. (1996). The reception of Mendeleev's periodic law in America and Britain. *Isis, 87*, 595–628.

Brush, S. G. (1999). Dynamics of theory change in chemistry: Part 2. Benzene and molecular orbitals, 1945–1980. *Studies in History and Philosophy of Science, 30*, 263–302.
Brush, S. G. (2000). Thomas Kuhn as a historian of science. *Science & Education, 9*, 39–58.
Bunce, D. M., & Robinson, W. R. (1997). Research in chemical education — The third branch of our profession. *Journal of Chemical Education, 74*(9), 1076–1079.
Bunge, M. (2003). Twenty-five centuries of quantum physics: From Pythagoras to us, and from subjectivism to realism. *Science & Education, 12*(5–6), 445–466.
Burbules, N. C., & Linn, M. C. (1991). Science education and philosophy of science: Congruence or contradiction? *International Journal of Science Education, 13*, 227–241.
Calver, N. (2013). Sir Peter Medawar: Science, creativity and the popularization of Karl Popper. *Notes and Records of the Royal Society, 67*, 301–314.
Campanario, J. M. (1993). Consolation for the scientist: Sometimes it is hard to publish papers that are later highly cited. *Social Studies of Science, 23*, 342–362.
Campanario, J. M. (1995). Commentary: On influential books and journal articles initially rejected because of negative referees' evaluations. *Science Communication, 16*, 304–325.
Campanario, J. M. (1996). Have referees rejected some of the most-cited articles of all times? *Journal of the American Society for Information Science, 47*, 302–310.
Campanario, J. M. (1998). Peer review as it stands today. Part I. *Science Communication, 19*, 181–211.
Campanario, J. M. (1999). La ciencia que no enseñamos. *Enseñanza de las Ciencias, 17*, 397–410.
Campanario, J. M. (2002). The parallelism between scientists' and students' resistance to new scientific ideas. *International Journal of Science Education, 24*, 1095–1110.
Campbell, D. T. (1988). Can we be scientific in applied social science? In E. S. Overman (Ed.), *Methodology and epistemology for social science* (pp. 315–333). Chicago: University of Chicago Press (first published 1984 in *Evaluation Studies Review Annual*), (pp. 315–333).
Cardellini, L. (2010). Modeling chemistry for effective chemical education: An interview with Ronald J Gillespie. *Journal of Chemical Education, 87*, 482–486.
Cardellini, L. (2013). Deep thinking. What are lectures useful for? *Journal of Chemical Education, 90*, 1418.
Carruthers, P., Stich, S., & Siegal, M. (Eds.). (2002). *The cognitive basis of science*. New York: Cambridge University Press.
Cartwright, N. (1983). *How the laws of physics lie*. Oxford: Clarendon Press.
Cartwright, N. (1989). *Nature's capacities and their measurement*. Oxford: Clarendon Press.
Cartwright, N. (1999). *The dappled world: A study of the boundaries of science*. Cambridge: Cambridge University Press.
Cavallo, A. (2008). Experiencing the nature of science: An interactive, beginning-of-semester activity. *Journal of College Science Teaching, 37*(5), 12–15. May/June.
Cha, D. (2007). *College physics*. Seoul: Books Hill.
Chalmers, A. (1998). Retracing the ancient steps to atomic theory. *Science & Education, 7*, 69–84.
Chalmers, A. (2009). *The scientist's atom and the philosopher's stone: How science succeeded and philosophy failed to gain knowledge of atoms*. Dordrecht: Springer.
Chalmers, A. (2010). Review symposium of the scientist's atom. *Metascience, 19*, 349–371.
Chang, R. (2007). *Chemistry* (9th ed.). New York: McGraw-Hill.
Cheong, Y. W., & Song, J. (2014). Different levels of the meaning of wave-particle duality and a suspensive perspective on the interpretation of quantum theory. *Science & Education, 23*(5), 1011–1030.
Chi, M. T. H. (2005). Commonsense conceptions of emergent processes. Why some misconceptions are robust. *The Journal of the Learning Sciences, 14*, 161–199.
Chinn, C. A., & Brewer, W. F. (1993). The role of anomalous data in knowledge acquisition: A theoretical framework and implications for science instruction. *Review of Educational Research, 63*, 1–49.

Christie, M. (1994). Philosophers versus chemists concerning 'laws of nature'. *Studies in History and Philosophy of Science, 25*, 613–629.
Clark, P. (1976). Atomism versus thermodynamics. In C. Howson (Ed.), *Method and appraisal in the physical sciences, 1800–1905* (pp. 41–105). Cambridge: Cambridge University Press.
Clement, J. (1998). Expert novice similarities and instruction using analogies. *International Journal of Science Education, 20*(10), 1271–1286.
Clough, M. P. (2006). Learners' responses to the demands of conceptual change: Considerations for effective native of science instruction. *Science & Education, 15*, 463–494.
Clough, M. P., & Olson, J. K. (2004). The nature of science: Always part of the science story. *The Science Teacher, 71*(9), 28–31.
Cobb, P., & Steffe, L. (1983). The constructivist researcher as theory and model builder. *Journal for Research in Mathematics Education, 14*, 83–94.
Cobern, W. W., & Loving, C. (2001). Defining 'science' in a multicultural world: Implications for science education. *Science Education, 85*, 50–67.
Cobern, W. W., Gibson, A. T., & Underwood, S. A. (1999). Conceptualizations of nature: An interpretative study of 16 ninth graders' everyday thinking. *Journal of Research in Science Teaching, 36*, 541–564.
Coffey, P. (2008). *Cathedrals of science: The personalities and rivalries that made modern chemistry*. New York: Oxford University Press.
Cohen, R. S. (1976). *Physical science*. New York: Holt, Rinehart and Winston.
Collins, H., & Pinch, T. (1993). *The golem: What everyone should know about science*. New York: Cambridge University Press.
Conant, J. B. (1947). *On understanding science: A historical approach*. Cambridge, MA: Harvard University Press.
Conant, J. B. (1948). *Education in a divided world*. Cambridge, MA: Harvard University Press.
Cooper, L. N. (1968). *An introduction to the meaning and structure of physics*. New York: Harper & Row.
Cooper, L. N. (1970). *An introduction to the meaning and structure of physics (short edition)*. New York: Harper & Row.
Cooper, L. N. (1992). *Physics: Structure and meaning*. Hanover: University Press of New England.
Cooper, M. M. (2013). Chemistry and the next generation science standards. *Journal of Chemical Education, 90*, 679–680.
Cooper, M. M., & Klymkowsky, M. (2013). Chemistry, life, the universe, and everything: A new approach to general chemistry, and a model for curriculum reform. *Journal of Chemical Education, 90*(9), 1116–1122.
Cooper, M. M., Grove, N., Underwood, S. M., & Klymkowsky, M. (2010). Lost in Lewis structures: An investigation of student difficulties in developing representational competence. *Journal of Chemical Education, 87*(8), 869–874.
Cooper, M. M., Underwood, S., Hilley, C. Z., & Klymkowsky, M. (2012). *Journal of Chemical Education, 89*, 1351–1357.
Coştu, B. (2007). Comparison of students' performance on algorithmic, conceptual and graphical chemistry gas problems. *Journal of Science Education and Technology, 16*, 379–386.
Cotes, S., & Cotuá, J. (2014). Using evidence response systems during interactive lectures to promote active learning and conceptual understanding of stoichiometry. *Journal of Chemical Education, 91*(5), 673–677.
Coulson, C. A. (1937). The evaluation of certain integrals occurring in studies of molecular structure. *Proceedings of the Cambridge Philosophical Society, 33*, 104–110.
Coulson, C. A. (1952). *Valence* (1st ed.). Oxford: Clarendon Press.
Coulson, C. A. (1961). *Valence* (2nd ed.). Oxford: Oxford University Press.
Coulson, C. A. (1970). Recent developments in valence theory. Symposium 'fifty years of valence theory'. *Pure and Applied Chemistry, 24*, 257–287.
Crease, R. M. (2002). Critical point: The most beautiful experiment. *Physics World, 15*(9), 19–20.

Croft, M., & De Berg, K. (2014). From common sense concepts to scientifically conditioned concepts of chemical bonding: An historical and textbook approach designed to address learning and teaching issues at the secondary school level. *Science & Education, 23*(9), 1733–1761.
Crowther, J. G. (1910). *Proceedings of the Royal Society* (Vol. lxxxiv). London: Royal Society.
Cushing, J. T. (1989). The justification and selection of scientific theories. *Synthese, 78*, 1–24.
Cushing, J. T. (1991). Quantum theory and explanatory discourse: Endgame for understanding. *Philosophy of Science, 58*, 337–358.
Cushing, J. T. (1994). *Quantum mechanics: historical contingency and the Copenhagen hegemony*. Chicago: University of Chicago Press: Chicago.
Cushing, J. T. (1996). The causal quantum theory program. In J. T. Cushing, A. Fine, & S. Goldstein (Eds.), *Bohmian mechanics and quantum theory: An appraisal* (Boston studies in the philosophy of science, Vol. 184, pp. 1–19). Dordrecht: Kluwer.
Cushing, J. T. (1998). *Philosophical concepts in physics: The historical relation between philosophy and scientific theories*. Cambridge: Cambridge University Press.
D'Ambrosio, B. S., & Campos, T. M. M. (1992). Preservice teachers' representation of children's understanding of mathematical concepts: Conflicts and conflict resolution. *Educational Studies in Mathematics, 23*, 213–230.
Dahsah, C., & Coll, R. K. (2007). Thai grade 10 and 11 students' conceptual understanding and ability to solve stoichiometry problems. *Research in Science and Technological Education, 25*(2), 227–241.
Darrigol, O. (2009). A simplified genesis of quantum mechanics. *Studies in History and Philosophy of Modern Physics, 40*, 151–166.
Daston, L., & Galison, P. (2007). *Objectivity*. New York: Zone Books.
Davisson, C., & Germer, L. H. (1927a). Diffraction of electrons by a crystal of nickel. *Physical Review, 30*(6), 705–740.
Davisson, C., & Germer, L. H. (1927b). The scattering of electrons by a single crystal of nickel. *Nature, 119*, 558.
Davisson, C., & Kunsman, C. H. (1923). The scattering of low speed electrons by, platinum and magnesium. *Physical Review, 22*, 242–258.
De Berg, K. C. (2003). The development of the theory of electrolytic dissociation: A case study of a scientific controversy and the changing nature of chemistry. *Science & Education, 12*, 397–419.
De Berg, K. C. (2006). The kinetic-molecular and thermodynamic approaches to osmotic pressure: A study of dispute in physical chemistry and its implications for chemistry education. *Science & Education, 15*(5), 495–519.
De Berg, K. C. (2014a). The place of history of chemistry in the teaching and learning of chemistry. In M. R. Matthews (Ed.), *International handbook of research in history, philosophy and science teaching* (pp. 317–341). Dordrecht: Springer.
De Berg, K. C. (2014b). The significance of the origin of physical chemistry for physical chemistry education: The case of electrolyte solution chemistry. *Chemistry Education Research and Practice, 15*, 266–275.
De Broglie, L. (1922). *Journal de Physique (Series VI), 3*, 422.
De Broglie, L. (1923a). Ondes et quanta. *Comptes Rendus, 177*, 507–510, 548–550, 630–632.
De Broglie, L. (1923b). Waves and quanta. *Nature, 112*, 540.
De Broglie, L. (1924). A tentative theory of light quanta. *Philosophical Magazine (Series 6), 47*, 446–458.
De Milt, C. (1952). The value of the history and philosophy of science in the training of graduate students in chemistry. *Journal of Chemical Education, 29*, 340–344.
De Souza, R. T., & Iyengar, S. S. (2013). Using quantum mechanics to facilitate the introduction of a broad range of chemical concepts to first-year undergraduate students. *Journal of Chemical Education, 90*, 717–725.
Demircioğlu, H., Demircioğlu, G., & Çalik, A. (2009). Investigating the effectiveness of storylines embedded within a context-based approach: The case for the periodic table. *Chemistry Education Research and Practice, 10*, 241–249.

DiGiuseppe, M. (2014). Representing nature of science in a science textbook: Exploring author-editor-publisher interactions. *International Journal of Science Education, 36*(7), 1061–1082.

Dirac, P. A. M. (1977). Ehrenhaft, the subelectron and the quark. In C. Weiner (Ed.), *History of twentieth century physics* (pp. 290–293). New York: Academic Press.

Drago, A. (2014). Il ruolo del sistema periodico degli elementi nel caratterizzare la chimica classica come teoria scientifica. *Epistemologia, 37*, 37–57.

Duhem, P. (1914). *The aim and structure of physical theory* (trans: Philip P. Wiener, 2nd ed.). New York: Atheneum.

Dunbar, R. E. (1938). Historical materials in college general chemistry textbooks. *Journal of Chemical Education, 15*, 183–186.

Duschl, R. A., & Duncan, R. G. (2009). Reply to both questions. In S. Tobias & T. M. Duffy (Eds.), *Constructivist instruction: Success or failure?* (p. 324). New York: Routledge.

Duschl, R. A., & Grandy, R. (2013). Two views about explicitly teaching nature of science. *Science & Education, 22*(9), 2109–2139.

Dyson, F. W., Eddington, A. S., & Davidson, C. (1920). A determination of the deflection of light by the sun's gravitational field, from observations made at the total eclipse of May 29, 1919. *Royal Society Philosophical Transactions, 220*, 291–333.

Earman, J. (2004). Laws, symmetry and symmetry breaking: Invariance, conservation principles and objectivity. *Philosophy of Science, 71*(5), 1227–1241.

Earman, J., & Glymour, C. (1980). Relativity and eclipses: The British eclipse expeditions of 1919 and their predecessors. *Historical Studies in the Physical Sciences, 11*(1), 49–85.

Eflin, J. T., Glennan, S., & Reisch, G. (1999). The nature of science: A perspective from the philosophy of science. *Journal of Research in Science Teaching, 36*(1), 107–116.

Ehenhaft, F. (1941). The microcoulomb experiment. *Philosophy of Science, 8*, 403–457.

Ehrenhaft, F. (1910). Uber die kleinsten messbaren elektrizitätsmengen. Zweite vorläufige mitteilung der methode zur bestimmung des elektrischen elementarquantums. *Anzeiger Akad. Wiss, 10*, 118–119 (Vienna).

Ehrenhaft, F. (1914). *Annalen der Physik, 44*, 657.

Einstein, A. (1905). Über einen Erzeugung und Verwandlung des Lichtes betreffenden heuristischen Gesichtspunkt. *Annalen der Physik, 17*, 132–148.

Einstein, A. (1909a). Zum gegenwärtigen stand des strahlungsproblems. *Physikalische Zeitschrift, 10*, 185–193.

Einstein, A. (1909b). Über die Entwicklung unsere anschauungen über das wesen und die konstitution der strahlung. *Physikaliche Zeitschrift, 10*, 817–825.

Einstein, A. (1916). Zur quantentheorie der strahlung. *Zürich Mitteilungen, 18*, 47–62.

El-Hani, C. N., Roque, N., & Rocha, P. L. B. (2005). *Programa Nacional do Livro Didático do Ensino Médio (PNLEM)*. Brasília: MEC.

Erduran, S. (2013). Philosophy, chemistry and education: An introduction. *Science & Education, 22*(7), 1559–1562.

Erduran, S., & Mugaloglu, E. Z. (2014). Philosophy of chemistry in chemical education: Recent trends and future directions. In M. R. Matthews (Ed.), *International handbook of research in history, philosophy and science teaching* (pp. 287–315). Dordrecht: Springer.

Falconer, I. (1987). Corpuscles, electrons, and cathode rays: J. J. Thomson and the 'discovery of the electron'. *British Journal for the History of Science, 20*, 241–276.

Farré, A. S., & Lorenzo, M. G. (2012). De la construcción del conocimiento científico a su enseñanza. Distintas explicaciones sobre la estructura del benceno. *Educacion Quimica, 23*, 271–279.

Feynman, R. (1967). *The character of physical law*. Cambridge, MA: MIT Press.

Fletcher, H. (1982). My work with Millikan on the oil-drop experiment. *Physics Today, 35*(6), 43–47.

Flick, L. B., & Lederman, N. G. (2004). Introduction. In L. B. Flick & N. G. Lederman (Eds.), *Scientific inquiry and nature of science* (pp. 9–18). Dordrecht: Kluwer.

Frank, J. O., & Lundsted, L. (1935). Historical materials in high school chemistry texts. *Journal of Chemical Education, 12*, 367–369.

Frické, M. (1976). The rejection of Avogadro's hypothesis. In C. Howson (Ed.), *Method and appraisal in the physical sciences: The critical background to modern science, 1800–1905* (pp. 277–307). Cambridge, UK: Cambridge University Press.

Furió-Más, C., Calatayud, M. L., Guisasola, J., & Furió-Gómez, C. (2005). How are concepts and theories of acid–base reactions presented? Chemistry in textbooks and as presented by teachers. *International Journal of Science Education, 27*(11), 1337–1358.

Gabel, D. L. (1993). Use of particle nature of matter in developing conceptual understanding. *Journal of Chemical Education, 70*, 193–194.

Gabel, D. L., & Bunce, D. M. (1994). Research on problem solving: Chemistry. In D. L. Gabel (Ed.), *Handbook of research on science teaching and learning* (pp. 301–326). New York: Macmillan.

Galison, P. (2008). Ten problems in history and philosophy of science. *Isis, 99*, 111–124.

Gallego Badillo, R., Gallego Torres, A. P., & Pérez Miranda, R. (2012). El congreso de Karlsruhe. Los inicios de una comunidad científica. *Educacion Quimica, 23*, 280–289.

Garritz, A. (2010). La historia como una herramienta para promover el aprendizaje. *Educacion Quimica, 21*(4), 266–268.

Garritz, A. (2013). Teaching the philosophical interpretations of quantum mechanics and quantum chemistry through controversies. *Science & Education, 22*, 1787–1807.

Garritz, A., Sosa, S., Hernández-Millán, G., López-Villa, N. M., Nieto-Calleja, E., Flor de María Reyes-Cárdenas, F., & César Robles Haro, C. (2013). Una secuencia de enseñanza/aprendizaje para los conceptos de sustancia y reacción química con base en la Naturaleza de la Ciencia y la Tecnología. *Educacion Quimica, 24*, 439–450.

Gavroglu, K. (1990). The reaction of the British physicists and chemists to van der Waal's early work and to the law of corresponding states. *Historical Studies in the Physical and Biological Sciences, 20*, 199–237.

Gavroglu, K. (2000). Controversies and the becoming of physical chemistry. In P. Machamer, M. Pera, & A. Baltas (Eds.), *Scientific controversies: Philosophical and historical perspectives* (pp. 177–198). New York: Oxford University Press.

Gavroglu, K., & Simões, A. (2012). *Neither physics nor chemistry: A history of quantum chemistry*. Cambridge, MA: Massachusetts Institute of Technology Press.

Gieger, H., & Marsden, E. (1909). On a diffuse reflection of the alpha particles. *Proceedings of the Royal Society* (Vol. lxxxii). London: Royal Society.

Giere, R. N. (1988). *Explaining science: A cognitive approach*. Chicago: University of Chicago Press.

Giere, R. N. (1995a). Viewing science. In R. Burian, D. Hull, & M. Forbes (Eds.), *PSA 1994* (Vol. 2). East Lansing: Philosophy of Science Association (revised version of the Presidential address at the Biennial Meeting of the Philosophy of Science Association, New Orleans, October, 1994).

Giere, R. N. (1995b). The skeptical perspective: Science without laws of nature. In F. Weinert (Ed.), *Laws of nature: Essays on the philosophical, scientific and historical dimensions* (pp. 120–138). Berlin: Walter de Gruyter.

Giere, R. N. (1999). *Science without laws*. Chicago: University of Chicago Press.

Giere, R. N. (2006a). *Scientific perspectivism*. Chicago: University of Chicago Press.

Giere, R. N. (2006b). Perspectival pluralism. In S. H. Kellert, H. E. Longino, & C. K. Waters (Eds.), *Scientific pluralism* (pp. 26–41). Minneapolis: University of Minnesota Press.

Giere, R. N. (2010). Naturalism. In S. Psillos & M. Curd (Eds.), *The Routledge companion to philosophy of science* (pp. 213–223). London: Routledge.

Giere, R. N. (2014). Email to author dated March 19, 2014 after reading a preliminary version of Chapter 2, reproduced with permission.

Gilbert, J. K. (Ed.). (2005). *Visualization in science education*. Dordrecht: Springer.

Gillespie, R. J. (1963). The valence-shell electron-pair repulsion (VSEPR) theory of directed valency. *Journal of Chemical Education, 40*, 295–299.

Gillespie, R. J. (1997a). Reforming the general chemistry textbook. *Journal of Chemical Education, 74*(5), 484–485.

Gillespie, R. J. (1997b). The great ideas of chemistry. *Journal of Chemical Education, 74*(7), 862–864.

Gillespie, R. J. (2008). Fifty years of the VSEPR model. *Coordination Chemistry Reviews, 252*, 1315–1327.

Gillespie, R. J., & Nyholm, R. S. (1957). Inorganic stereochemistry. *Quarterly Review of the Chemical Society, 11*, 339–380.

Gillespie, R. J., Spencer, J. N., & Moog, R. S. (1996). Part 2. Bonding and molecular geometry without orbitals — The electron domain model. *Journal of Chemical Education, 73*, 622–627.

Good, R. (1993). Editorial: The many forms of constructivism. *Journal of Research in Science Teaching, 30*, 1015.

Gooday, G., Lynch, J. M., Wilson, K. G., & Barsky, C. K. (2008). Does science education need the history of science? *Isis, 99*, 322–330.

Gordin, M. D. (2004). *A well-ordered thing: Dmitrii Mendeleev and the shadow of the periodic table*. New York: Basic Books.

Gordin, M. (2012). The textbook case of a priority dispute: D.I. Mendeleev, Lothar Meyer, and the periodic system. In M. Biagioli & J. Riskin (Eds.), *Nature engaged: Science in practice from the renaissance to the present* (pp. 59–82). New York: Palgrave Macmillan.

Greca, I. M., & Freire, O., Jr. (2014). Teaching introductory quantum physics and chemistry: Caveats from the history of science and science teaching to the training of modern chemists. *Chemistry Education Research and Practice, 15*, 286–296.

Gulacar, O., Eilks, I., & Bowman, C. R. (2014). Differences in general cognitive abilities and domain-specific skills of higher– and lower-achieving students in stoichiometry. *Journal of Chemical Education, 91*, 961–968.

Gultepe, N., Celik, A. Y., & Kilic, Z. (2013). Exploring effects of high school students' mathematical processing skills and conceptual understanding of chemical concepts on algorithmic problem solving. *Australian Journal of Teacher Education, 38*(10), 105–122.

Hanson, N. R. (1958). *Patterns of discovery*. Cambridge: Cambridge University Press.

Heering, P., & Klassen, S. (2010). Doing it differently: Attempts to improve oil-drop experiment. *Physics Education, 45*(4), 382–393.

Heering, P., & Klassen, S. (2011). Troublesome droplets: Improving students' experiences with the Millikan oil drop experiment. In P. V. Kokkotas, K. S. Malamitsa, & A. A. Rizaki (Eds.), *Adapting historical knowledge production to the classroom* (pp. 103–112). Rotterdam: Sense Publishers.

Heilbron, J. L. (1964). A history of atomic models from the discovery of the electron to the beginnings of quantum mechanics. Doctoral dissertation, University of California, Berkeley.

Heilbron, J. L. (1981). *Historical studies in the theory of atomic structure*. New York: Arno Press.

Heilbron, J. L., & Kuhn, T. (1969). The genesis of the Bohr atom. *Historical Studies in the Physical Sciences, 1*, 211–290.

Herron, J. D. (1977). Rutherford and the nuclear atom. *Journal of Chemical Education, 54*, 499.

Heyrovska, R. (1996). Physical electrochemistry of strong electrolytes based on partial dissociation and hydration. *Journal of the Electrochemical Society, 143*(6), 1789–1793.

Hodson, D. (2009). *Teaching and learning about science: Language, theories, methods, history, traditions and values*. Rotterdam: Sense Publishers.

Hodson, D., & Wong, S. L. (2014). From the horse's mouth: Why scientists' views are crucial to nature of science understanding. *International Journal of Science Education, 36*(16), 2639–2665.

Hoffmann, R. (2012). In J. Kovac & M. Weisberg (Eds.), *Roald Hoffmann on the philosophy, art, and science of chemistry*. New York: Oxford University Press.

Hoffmann, R., Shaik, S., & Hiberty, P. C. (2003). A conversation on VB vs MO theory: A never-ending rivalry? *Accounts of Chemical Research, 36*(10), 750–756.

Holton, G. (1952). *Introduction to concepts and theories in physical science*. New York: Addison-Wesley.

Holton, G. (1969). Einstein and the 'crucial' experiment. *American Journal of Physics, 37*, 968–982.

Holton, G. (1978a). Subelectrons, presuppositions, and the Millikan-Ehrenhaft dispute. *Historical Studies in the Physical Sciences, 9*, 161–224.

Holton, G. (1978b). *The scientific imagination: Case studies*. Cambridge: Cambridge University Press.

Holton, G. (1986). *The advancement of science and its burdens*. Cambridge: Cambridge University Press.

Holton, G. (1988). On the hesitant rise of quantum physics research in the United States. In S. Goldberg & R. H. Stuewer (Eds.), *The Michelson Era in American science, 1970–1930* (pp. 177–205). New York: American Institute of Physics.

Holton, G. (1998). *The scientific imagination*. Cambridge, MA: Harvard University Press.

Holton, G. (1999). R.A. Millikan's struggle with the meaning of Planck's constant. *Physics in Perspective, 1*, 231–237.

Holton, G. (2014a). The neglected mandate: Teaching science as part of our culture. *Science & Education, 23*, 1875–1877.

Holton, G. (2014b). Personal communication to the author, August 3, italics in the original. Reproduced with permission.

Holton, G., & Brush, S. G. (2001). *Physics, the human adventure: From Copernicus to Einstein and beyond*. New Brunswick: Rutgers University Press.

Höttecke, D., Henke, A., & Riess, F. (2012). Implementing history and philosophy in science teaching: Strategies, methods, results and experiences from the European HIPST project. *Science & Education, 21*, 1233–1261.

Howson, C. (Ed.). (1976). *Method and appraisal in the physical sciences: The critical background to modern sciences, 1800–1905*. Cambridge: Cambridge University Press.

Howson, C. (1990). Essay review: The poverty of historicism. *Studies in History and Philosophy of Science, 21*, 173–179.

Hoyningen-Huene, P. (2013). *Systematicity: The nature of science*. Oxford, UK: Oxford University Press.

Irwin, A. R. (2000). Historical case studies: Teaching the nature of science in context. *Science Education, 84*, 5–26.

Irzik, G., & Nola, R. (2011). A family resemblance approach to the nature of science for science education. *Science & Education, 20*(7–8), 591–607.

Jaffe, B. (1938). The history of chemistry and its place in the teaching of high school chemistry. *Journal of Chemical Education, 15*, 383–389.

Jenkins, E. (2007). School science: a questionable construct? *Journal of Curriculum Studies, 39*(3), 265–282.

Jensen, W. B. (1984). Abegg, Lewis, Langmuir, and the octet rule. *Journal of Chemical Education, 61*, 191–200.

Johnstone, A. H. (1991). Thinking about thinking. *International Newsletter on Chemical Education, 36*, 7–11.

Jones, R. C. (1995). The Millikan oil-drop experiment: Making it worthwhile. *American Journal of Physics, 63*, 970–977.

Justi, R., & Gilbert, J. (2000). History and philosophy of science through models: Some challenges in the case of 'the atom'. *International Journal of Science Education, 22*, 993–1009.

Justi, R., & Gilbert, J. (1999). A cause of ahistorical science teaching: Use of hybrid models. *Science Education, 83*, 163–177

Kang, S., Scharmann, L. C., & Noh, T. (2005). Examining students' views on the nature of science: Results from Korean 6th, 8th, and 10th graders. *Science Education, 89*, 314–334.

Kaufman, W. (1897). Die magnetische ablenkbarkeit der kathodenstrahlen und ihre abhängigkeit vom entladungspotential. *Annalen der Physik und Chemie, 61*, 544.

Kellert, S. H., Longino, H. E., & Waters, C. K. (2006). Introduction: The pluralist stance. In S. H. Kellert, H. E. Longino, & C. K. Waters (Eds.), *Scientific pluralism* (Minnesota studies in the philosophy of science, Vol. XIX, pp. 7–29). Minneapolis: University of Minnesota Press.

Khishfe, R., & Lederman, N. G. (2007). Relationship between instructional context and views of nature of science. *International Journal of Science Education, 29*, 939–961.

Kirschner, P. A., Sweller, J., & Clark, R. (2006). Why minimal guidance during instruction does not work: An analysis of the failure of constructivist, discovery, problem-based, experiential and inquiry-based teaching. *Educational Psychologist, 41*, 75–86.

Kitchener, R. F. (1986). *Piaget's theory of knowledge: Genetic epistemology and scientific reason*. New Haven: Yale University Press.

Kitchener, R. F. (1993). Piaget's epistemic subject and science education: Epistemological vs psychological issues. *Science & Education, 2*, 137–148.

Kitcher, P. (2007). *Living with Darwin: Evolution, design, and the future of faith*. New York: Oxford University Press.

Kitcher, P. (2011). Epistemology without history is blind. *Erkenntnis, 75*, 505–524.

Klahr, D. (2009). Question: Klahr. In S. Tobias & T. M. Duffy (Eds.), *Constructivist instruction: Success or failure?* (p. 323). New York: Routledge.

Klassen, S. (2006). A theoretical framework for contextual science teaching. *Interchange, 37*, 31–62.

Klassen, S. (2009). Identifying and addressing student difficulties with the Millikan oil drop experiment. *Science & Education, 18*, 593–607.

Klassen, S., Niaz, M., Metz, D., McMillan, B., & Dietrich, S. (2012). Portrayal of the history of the photoelectric effect in laboratory instructions. *Science & Education, 21*, 729–743.

Klein, U. (2003). *Experiments, models, paper tools: Cultures of organic chemistry in the nineteenth century*. Stanford, CA: Stanford University Press.

Klopfer, L. E. (1969). The teaching of science and the history of science. *Journal of Research in Science Teaching, 6*, 87–95.

Kohler, R. E. (1971). The origin of Lewis's theory of the shared pair bond. *Historical Studies in the Physical Sciences, 3*, 343–376.

Kolstø, S. D. (2008). Science education for democratic citizenship through the use of history of science. *Science & Education, 17*, 977–997.

Kotz, J. C., & Purcell, K. (1991). *Chemistry and chemical reactivity* (2nd ed.). Philadelphia: Saunders.

Kousathana, M., Demerouti, M., & Tsaparlis, G. (2005). Instructional misconceptions in acid–base equilibria: A analysis from a history and philosophy of science perspective. *Science & Education, 14*, 173–194.

Krane, K. S. (1996). *Modern physics* (2nd ed.). New York: Wiley.

Kuhn, T. S. (1962). *The structure of scientific revolutions*. Chicago: University of Chicago Press.

Kuhn, T. S. (1970). *The structure of scientific revolutions* (2nd ed.). Chicago: University of Chicago Press.

Kuhn, T. S. (1977). The function of measurement in modern physical science. In T. S. Kuhn (Ed.), *Essential tension* (pp. 178–224). Chicago: University of Chicago Press (originally published in *Isis, 52*, 161–190, 1961).

Kuhn, T. S. (1978). *Black-body theory and quantum discontinuity: 1894–1912*. New York: Oxford University Press.

Labarca, M., Bejarano, N., & Eichler, M. L. (2013). Química e filosofía: Rumo a una frutífera colaboração. *Quimica Nova, 36*(8), 1256–1266.

Lakatos, I. (1970). Falsification and the methodology of scientific research programmes. In I. Lakatos & A. Musgrave (Eds.), *Criticism and the growth of knowledge* (pp. 91–195). Cambridge: Cambridge University Press.

Lakatos, I. (1971). History of science and its rational reconstructions. In R. C. Buck & R. S. Cohen (Eds.), *Boston studies in the philosophy of science* (Vol. 8, pp. 91–136). Dordrecht: Reidel.

Laloë, F. (2001). Do we really understand quantum mechanics? Strange correlations, paradoxes, and theorems. *American Journal of Physics, 69*, 655–701.

Landau, L. D., & Lifshitz, E. M. (1958). *Quantum mechanics: Non-relativistic theory*. Reading: Addison-Wesley.

Laudan, L. (1977). *Progress and its problems: Towards a theory of scientific growth*. Berkeley: University of California Press.

Laudan, L. (1981). A confutation of convergent realism. *Philosophy of Science, 48*, 33–49.

Laudan, L. (1996). *Beyond positivism and relativism: Theory, method and evidence*. Boulder: Westview Press (Division of HarperCollins).

Laudan, L., Donovan, A., Laudan, R., Barker, P., Brown, H., Leplin, J., Thagard, P., & Wykstra, S. (1986). Scientific change: Philosophical models and historical research. *Synthese, 69*, 141–223.

Laudan, R., Laudan, L., & Donovan, A. (1988). Testing theories of scientific change. In A. Donovan, L. Laudan, & R. Laudan (Eds.), *Scrutinizing science: Empirical studies of scientific change* (pp. 3–44). Dordrecht: Kluwer.

Lautesse, P., Valls, A. V., Ferlin, F., Héraud, J.-L., & Chabot, H. (2015). Teaching quantum physics in upper secondary school in France: 'Quanton' versus 'wave-particle' duality, two approaches of the problem of reference. *Science & Education, 24*, 937–955.

Lawler, J. (1975). Dialectical philosophy and developmental psychology: Hegel and Piaget on contradiction. *Human Development, 18*, 1–17.

Lawson, A. E. (1983). Predicting science achievement: The role of developmental level, disembedding level, mental capacity, prior knowledge, and beliefs. *Journal of Research in Science Teaching, 20*, 117–129.

Lawson, A. E. (1985). A review of research on formal reasoning and science instruction. *Journal of Research in Science Teaching, 22*, 569–617.

Lawson, A. E., Abraham, M. R., & Renner, J. W. (1989). A theory of instruction: Using the learning cycle to teach science concepts and thinking skills. *Monographs of the National Association for Research in Science Teaching, 1*, 1–57.

Lederman, N. G. (1992). Students' and teachers' conceptions of the nature of science: A review of the research. *Journal of Research in Science Teaching, 29*, 331–359.

Lederman, N. G. (2004). Syntax of nature of science within inquiry and science instruction. In L. B. Flick & N. G. Lederman (Eds.), *Scientific inquiry and nature of science* (pp. 301–317). Dordrecht: Kluwer.

Lederman, N. G., & Abd-El-Khalick, F. (1998). Avoiding de-natured science: Activities that promote understanding of the nature of science. In W. F. McComas (Ed.), *The nature of science in science education: Rationales and strategies* (pp. 83–126). Dordrecht: Kluwer Academic Publishers.

Lederman, N. G., & Lederman, J. S. (2013). Next generation science teacher educators. *Journal of Science Teacher Education, 24*, 929–932.

Lederman, N. G., Abd-El-Khalick, F., Bell, R. L., & Schwartz, R. (2002). Views of nature of science questionnaire: Toward valid and meaningful assessment of learners' conceptions of nature of science. *Journal of Research in Science Teaching, 39*, 497–521.

Lee, G., & Yi, J. (2013). Where cognitive conflict arises from? The structure of creating cognitive conflict. *International Journal of Science and Mathematics Education, 11*, 601–623.

Lee, M.-H., Wu, Y.-T., & Tsai, C.-C. (2009). Research trends in science education from 2003 to 2007: A content analysis of publications in selected journals. *International Journal of Science Education, 31*, 1999–2020.

Lehrer, R., & Schauble, L. (2012). Seeding evolutionary by engaging children in modeling its foundations. *Science Education, 96*(4), 701–724.

Leite, L. (2002). History of science in science education: Development and validation of a checklist for analyzing the historical of science textbooks. *Science & Education, 11*, 333–359.

Lenard, P. E. A. (1902). Über die LichtelektrischeWirkung. *Annalen der Physik, 8*, 149–198.

Levere, T. H. (2006). What history can teach us about science: Theory and experiment, data and evidence. *Interchange, 37*, 115–128.

Lévy-Leblond, J. M. (1988). Neither waves, nor particles, but quantons. *Nature, 334*(6177), 19–20.

Lewis, G. N. (1916). The atom and the molecule. *Journal of American Chemical Society, 38*, 762–785.

Lewis, G. N. (1923). *Valence and the structure of atoms and molecules*. New York: Chemical Catalog.

Lipton, P. (2005a). Testing hypotheses: Prediction and prejudice. *Science, 307*(14 January), 219–221.

Lipton, P. (2005b). Response. *Science, 308*, 1411–1412.

Lombardi, O., & Labarca, M. (2007). The philosophy of chemistry as a new resource for chemistry education. *Journal of Chemical Education, 84*(1), 187–192.

Longino, H. E. (1990). *Science as social knowledge*. Princeton: Princeton University Press.

Lorenzelli, V. (1969). *Chimica. Per student di ingegneria*. Genova: Edizioni Culturali Italiane.

Losee, J. (2001). *A historical introduction to the philosophy of science* (4th ed.). Oxford: Oxford University Press.

Lynch, S., & Bryan, L. (2014). Supporting the implementation of the next generation science standards (NGSS) through research: Introduction to NSRST position papers. https://narst.org/ngss-papers/. Retrieved on August 6, 2014.

Machamer, P., & Wolters, G. (2004). Introduction. In P. Machamer & G. Wolters (Eds.), *Science, values, and objectivity* (pp. 1–13). Pittsburgh: University of Pittsburgh Press.

Mahan, B., & Myers, R. J. (1990). *University chemistry* (4th ed.). Menlo Park: Benjamin Cummings. Spanish.

Maher, P. (1988). Prediction, accommodation and the logic of discovery. In A. Fine & J. Leplin (Eds.), *PSA 1988* (Vol. 1). East Lansing: Philosophy of Science Association.

Margenau, H. (1950). *The nature of physical reality*. New York: McGraw-Hill.

Matthews, M. R. (1987). Experiment as the objectification of theory: Galileo's revolution. Proceedings of the second international seminar on misconceptions and educational strategies in science and mathematics (I, pp. 289–298). Ithaca: Cornell University.

Matthews, M. R. (1994). *Science teaching: The contribution of history and philosophy of science*. New York: Routledge.

Matthews, M. R. (1997). Introductory comments on philosophy and constructivism in science education. *Science & Education, 6*, 5–14.

Matthews, M. R. (1998). In defense of modest goals in when teaching about the nature of science. *Journal of Research in Science Teaching, 35*, 161–174.

Matthews, M. R. (2012). Changing the focus: From nature of science to features of science. In M. S. Khine (Ed.), *Advances in nature of science research* (pp. 3–26). Dordrecht: Springer.

Matthews, M. R. (2015). *Science teaching: The contribution of history and philosophy of science* (20th Anniversary revised and expanded edition). New York: Routledge.

Maxwell, J. C. (1860). Illustrations of the dynamical theory of gases. *Philosophical Magazine, 19*, 19–32 (Reproduced in Scientific Papers, 1965, pp. 377–409, New York: Dover).

McComas, W. F., & Olson, J. K. (1998). The nature of science in international science education standards documents. In W. F. McComas (Ed.), *The nature of science in science education: Rationales and strategies* (pp. 41–53). Dordrecht: Kluwer.

McComas, W. F., Clough, M. P., & Almazroa, H. (1998). The role and character of the nature of science in science education. In W. F. McComas (Ed.), *The nature of science in science education: Rationales and strategies* (pp. 3–40). Dordrecht: Kluwer.

McMullin, E. (1985). Galilean idealization. *Studies in History and Philosophy of Science, 16*, 247–273.

Medawar, P. B. (1967). *The art of the soluble*. London: Methuen.

Medicus, H. A. (1974). Fifty years of matter waves. *Physics Today, 27*, 38–45.

Mehlecke, C., Eichler, M. L., Miskinis Salgado, T. D., & Claudio Pino del, J., (2012). A abordagem histórica acerca de produção e da recepção da tabela periódica en libros didácticos brasileiros para o ensino médio. *Revista Electrónica de Enseñanza de las Ciencias, 11*(3), 521–545.

Mendeleev, D. (1879). The periodic law of the chemical elements. *The Chemical News, 40*. No. 1042.

Mendeleev, D. (1889). The periodic law of the chemical elements. *Journal of the Chemical Society, 55*, 634–656. Faraday lecture, delivered on 4 June 1889.

Mendeleev, D. (1897). *The principles of chemistry* (2nd English ed., trans of 6th Russian ed.). New York: American Home Library Company.
Millikan, R. A. (1913). On the elementary electrical charge and the Avogadro constant. *Physical Review, 2*, 109–143.
Millikan, R. A. (1916). A direct photoelectric determination of Planck's 'h'. *Physical Review, 7*, 355–388.
Millikan, R. A. (1917). *The electron: Its isolation and measurement and the determination of some of its properties*. Chicago: University of Chicago Press.
Millikan, R. A. (1947). *Electrons (+ and -), protons, photons, neutrons, mesotrons, and cosmic rays* (2nd ed.). Chicago: University of Chicago Press. first published 1935.
Millikan, R. A. (1950). *The autobiography of Robert A. Millikan*. Englewood Cliffs: Prentice-Hall, Inc.
Millikan, R. A. (1965). The electron and the light-quant from the experimental point of view. In *Nobel Lectures: Physics* (Nobel prize acceptance speech, 1923). Amsterdam: Elsevier.
Monk, M., & Osborne, J. (1997). Placing the history and philosophy of science on the curriculum: A model for the development of pedagogy. *Science Education, 81*, 405–424.
Moore, J. W. (2003). Editorial: Turning the (periodic) tables. *Journal of Chemical Education, 80*(8), 847.
Moore, J. W., Stanitski, C. L., & Jurs, P. C. (2011). *Chemistry: The molecular science* (4th edn.). Belmont: Brooks/Cole.
Morrison, M., & Morgan, M. S. (1999). Models as mediating instruments. In M. S. Morgan & M. Morrison (Eds.), *Models as mediators: Perspectives on natural and social science* (pp. 10–37). Cambridge, UK: Cambridge University Press.
Mortimer, E. F., & Machado, A. H. (2005). *Química*. São Paulo: Scipione.
Moseley, H. G. J. (1913–1914). Atomic models and X-ray spectra. *Nature, 92*, 554.
Mount Holyoke College. (2004). http://mtholyoke.edu/acad/intdept/i42/Photoeleclab04.PDF
Mulliken, R. S. (1932). Electronic structure of polyatomic molecules and valence. *Physical Review, 40*, 55–62.
National Research Council, NRC. (1996). *National science education standards*. Washington, DC: National Academy Press.
National Research Council, NRC. (2007). Taking science to school: Learning and teaching science kindergarten to eighth grade. In R. A. Duschl, H. A. Schweingruber, & A. W. Shouse (Eds.), *Center for education, division of behavioral and social sciences*. Washington, DC: The National Academy Press.
National Research Council, NRC. (2012). *Discipline-based education research: Understanding and improving learning in undergraduate science and engineering*. Washington, DC: The National Academies Press.
National Research Council, NRC. (2013). Next Generation Science Standards (NGSS). Washington, DC: National Academies Press. http://www.nextgenscience.org
Navarro, J. (2012). *A history of the electron: J.J. Thomson and G.P. Thomson*. Cambridge: Cambridge University Press.
Needham, P. (2004a). When did atoms begin to do any explanatory work in chemistry? *International Studies in the Philosophy of Science, 18*, 199–219.
Needham, P. (2004b). Has Daltonian atomism provided chemistry with any explanations? *Philosophy of Science, 71*, 1038–1047.
Needham, P. (2010). Review symposium of Alan Chalmers's the scientist's atom. *Metascience, 19*, 349–371.
Neressian, N. (2002). The cognitive basis of model-based reasoning in science. In P. Carruthers, S. Stich, & M. Siegel (Eds.), *The cognitive basis of science* (pp. 133–153). Cambridge: Cambridge University Press.
Newton-Smith, W. H. (1981). *The rationality of science*. London: Routledge and Kegan Paul.
Niaz, M. (1988). Manipulation of M-demand of chemistry problems and its effect on student performance. A neo-Piagetian study. *Journal of Research in Science Teaching, 25*, 643–657.

Niaz, M. (1989). The relationship between M-demand, algorithms, and problem solving: A neo-Piagetian analysis. *Journal of Chemical Education, 66*, 422–424.
Niaz, M. (1991). Role of the epistemic subject in Piaget's genetic epistemology and its importance for science education. *Journal of Research in Science Teaching, 28*, 569–580.
Niaz, M. (1992). From Piaget's epistemic subject to Pascual-Leone's metasubject: Epistemic transition in the constructivist-rationalist theory of cognitive development. *International Journal of Psychology, 27*(6), 443–457.
Niaz, M. (1993). If Piaget's epistemic subject is dead, shall we bury the scientific research methodology of idealization. *Journal of Research in Science Teaching, 30*(7), 809–812.
Niaz, M. (1994). Enhancing thinking skills: Domain specific/domain general strategies—A dilemma for science education. *Instructional Science, 22*, 413–422.
Niaz, M. (1995a). Cognitive conflict as a teaching strategy in solving chemistry problems: A dialectic-constructivist perspective. *Journal of Research in Science Teaching, 32*, 959–970.
Niaz, M. (1995b). Progressive transitions from algorithmic to conceptual understanding in student ability to solve chemistry problems: A Lakatosian interpretation. *Science Education, 79*, 19–36.
Niaz, M. (1995c). Piaget's epistemic subject: A reply to Shayer. *Journal of Research in Science Teaching, 32*, 1003–1005.
Niaz, M. (1997). How early can children understand some form of scientific reasoning? *Perceptual and Motor Skills, 85*, 1272–1274.
Niaz, M. (1998a). A Lakatosian conceptual change teaching strategy based on student ability to build models with varying degrees of conceptual understanding of chemical equilibrium. *Science & Education, 7*, 107–127.
Niaz, M. (1998b). From cathode rays to alpha particles to quantum of action: A rational reconstruction of structure of the atom and its implications for chemistry textbooks. *Science Education, 82*, 527–552.
Niaz, M. (1999). Should we put observations first? *Journal of Chemical Education, 75*, 734.
Niaz, M. (2000a). The oil drop experiment: A rational reconstruction of the Millikan-Ehrenhaft controversy and its implications for chemistry textbooks. *Journal of Research in Science Teaching, 37*, 480508.
Niaz, M. (2000b). A framework to understand students' differentiation between heat energy and temperature and its educational implications. *Interchange, 31*(1), 1–20.
Niaz, M. (2000c). A rational reconstruction of the kinetic molecular theory of gases based on history and philosophy of science and its implications for chemistry textbooks. *Instructional Science, 28*, 23–50.
Niaz, M. (2001a). Understanding nature of science as progressive transitions in heuristic principles. *Science Education, 85*, 684–690.
Niaz, M. (2001b). How important are the laws of definite and multiple proportions in chemistry and teaching chemistry? — A history and philosophy of science perspective. *Science & Education, 10*, 243–266.
Niaz, M. (2001c). A rational reconstruction of the origin of the covalent bond and its implications for general chemistry textbooks. *International Journal of Science Education, 23*(6), 623–641.
Niaz, M. (2005). An appraisal of the controversial nature of the oil drop experiment: Is closure possible? *British Journal for the Philosophy of Science, 56*(4), 681–702.
Niaz, M. (2008). *Teaching general chemistry: A history and philosophy of science approach*. New York: Nova Science Publishers.
Niaz, M. (2009). *Critical appraisal of physical science as a human enterprise: Dynamics of scientific progress*. Dordrecht: Springer.
Niaz, M. (2010). Are we teaching science as practiced by scientists? *American Journal of Physics, 78*, 5–6.
Niaz, M. (2011). *Innovating science teacher education: A history and philosophy of science perspective*. New York: Routledge.
Niaz, M. (2012a). *From 'science in the making' to understanding the nature of science: An overview for science educators*. New York: Routledge.

Niaz, M. (2012b). Editorial: Filosofía de la química o filosofía de la ciencia como guía para comprender el desarrollo de la química. *Educacion Quimica, 23*, 244–247.

Niaz, M. (2013). Kostas Gavroglu and Ana Simões: Neither physics nor chemistry: A history of quantum chemistry (Review). *Science & Education, 22*(3), 753–758.

Niaz, M. (2014a). Science textbooks: The role of history and philosophy of science. In M. R. Matthhews (Ed.), *International handbook of research in history, philosophy and science teaching* (pp. 1411–1441). Dordrecht: Springer.

Niaz, M. (2014b). Evaluation of textbooks: Approaches and consequences. In R. Gunstone (Ed.), *Encyclopedia of science education* (pp. 403–406). Dordrecht: Springer.

Niaz, M., & Cardellini, L. (2011a). What can the Bohr-Sommerfeld model show students of chemistry in the 21st century? *Journal of Chemical Education, 88*, 240–243.

Niaz, M., & Cardellini, L. (2011b). Why has the Bohr-Sommerfeld model of the atom been ignored by general chemistry textbooks? *Acta Chimica Slovenica, 58*, 876–883.

Niaz, M., & Coştu, B. (2009). Presentation of atomic structure in Turkish general chemistry textbooks. *Chemistry Education Research and Practice, 10*, 233–240.

Niaz, M., & Coştu, B. (2013). Analysis of Turkish general chemistry textbooks based on a history and philosophy of science perspective. In M. S. Khine (Ed.), *Critical analysis of science textbooks: Evaluating instructional effectiveness* (pp. 199–218). Dordrecht: Springer.

Niaz, M., & Fernández, R. (2008). Understanding quantum numbers in general chemistry textbooks. *International Journal of Science Education, 30*(7), 869–901.

Niaz, M., & Lawson, A. E. (1985). Balancing chemical equations: The role of developmental level and mental capacity. *Journal of Research in Science Teaching, 22*, 41–51.

Niaz, M., & Luiggi, M. (2014). *Facilitating conceptual change in students' conceptual understanding of the periodic table*. Dordrecht: Springer.

Niaz, M., & Marcano, C. (2012). *Reconstruction of wave-particle duality and its implications for general chemistry textbooks*. Dordrecht: Springer.

Niaz, M., & Maza, A. (2011). *Nature of science in general chemistry textbooks*. Dordrecht: Springer.

Niaz, M., & Montes, L. A. (2012). Understanding stoichiometry: Towards a history and philosophy of chemistry. *Educacion Quimica, 23*, 290–297.

Niaz, M., & Robinson, W. R. (1992). Manipulation of logical structure of chemistry problems and its effects on student performance. *Journal of Research in Science Teaching, 29*, 211–226.

Niaz, M., & Robinson, W. R. (1993). Teaching algorithmic problem solving or conceptual understanding: Role of developmental level, mental capacity, and cognitive style. *Journal of Science Education and Technology, 2*, 407–416.

Niaz, M., & Rodríguez, M. A. (2001). Do we have to introduce history and philosophy of science or is it already 'inside' chemistry? *Chemistry Education: Research and Practice in Europe, 2*, 159–164.

Niaz, M., Aguilera, D., Maza, A., & Liendo, G. (2002). Arguments, contradictions, resistances and conceptual change in students' understanding of atomic structure. *Science Education, 86*, 505–525.

Niaz, M., Abd-El-Khalick, F., Benarroch, A., Cardellini, L., Laburú, C. E., Marín, N., Montes, L. A., Nola, R., Orlik, Y., Scharmann, L. C., Tsai, C.-C., & Tsaparlis, G. (2003). Constructivism: Defense or a continual critical appraisal — A response to Gil-Pérez et al. *Science & Education, 12*, 787–797.

Niaz, M., Rodríguez, M. A., & Brito, A. (2004). An appraisal of Mendeleev's contribution to the development of the periodic table. *Studies in the History and Philosophy of Science, 35*, 271–282.

Niaz, M., Klassen, S., McMillan, B., & Metz, D. (2010a). Reconstruction of the history of the photoelectric effect and its implications for general physics textbooks. *Science Education, 94*, 903–931.

Niaz, M., Klassen, S., McMillan, B., & Metz, D. (2010b). Leon Cooper's perspective on teaching science: An interview study. *Science & Education, 19*, 39–54.

Niaz, M., Kwon, S., Kim, N., & Lee, G. (2013). Do general physics textbooks discuss scientists' ideas about atomic structure? A case in Korea. *Physics Education, 48*(1), 57–64.

Nola, R. (1997). Constructivism in science and science education: A philosophical critique. *Science & Education, 6*, 55–83.

Novak, J. D. (1990). Concept mapping: A useful tool for science education. *Journal of Research in Science Teaching, 27*(10), 937–949.

Nurrenbern, S. C., & Pickering, M. (1987). Concept learning versus problem solving: Is there a difference? *Journal of Chemical Education, 64*, 508–510.

Ogilivie, J. F. (1990). The nature of the chemical bond — 1990: There are no such things as orbitals! *Journal of Chemical Education, 67*, 280–289.

Oh, J.-Y. (2011). Using an enhanced conflict map in the classroom (photoelectric effect) based on Lakatosian heuristic principle strategies. *International Journal of Science and Mathematics Education, 9*, 1135–1166.

Osborne, J. F. (1996). Beyond constructivism. *Science Education, 80*, 53–82.

Osborne, J., Collins, S., Ratcliffe, M., Millar, R., & Duschl, R. (2003). What 'ideas-about-science' should be taught in school science? A Delphi study of the expert community. *Journal of Research in Science Teaching, 40*(7), 692–720.

Ospina, J. (2010). *Efecto fotoeléctrico: Una reconstrucción racional basada en la historia y filosofía de la ciencia y sus implicaciones para los textos de química general* (Master of Science (chemistry education) dissertation). Cumaná: Universidad de Oriente.

Ostwald, W. (1908). *Grundriss der Naturphilosophie* (2nd ed.). Leipzig: Philip Reclam.

Padilla, K., & Furio-Mas, C. (2008). The importance of history and philosophy of science in correcting distorted views of 'amount of substance' and 'mole' concepts in chemistry teaching. *Science & Education, 17*, 403–424.

Padilla, K., & Van Driel, J. H. (2011). The relationships between PCK components: The case of quantum chemistry professors. *Chemistry Education Research and Practice, 12*, 367–378.

Páez, Y., Rodríguez, M. A., & Niaz, M. (2004). Los modelos atómicos desde la perspectiva de la historia y filosofía de la ciencia: Un análisis de la imagen reflejada por los textos de química de bachillerato. *Investigación y Postgrado, 19*(1), 51–77.

Paraskevopoulou, E., & Koliopoulos, D. (2011). Teaching nature of science through the Millikan-Ehrenhaft dispute. *Science & Education, 20*(10), 943–950.

Pascual-Leone, J. (1970). A mathematical model for the transition rule in Piaget's developmental stages. *Acta Psychologica, 32*, 301–345.

Pascual-Leone, J. (1976). A view of cognition from a formalist's perspective. In K. F. Riegel & J. A. Meacham (Eds.), *The developing individual in a changing world* (pp. 89–100). The Hague: Mouton.

Pascual-Leone, J. (1978). Compounds, confounds, and models in developmental information processing: A reply to Trabasso and Foellinger. *Journal of Experimental Child Psychology, 26*, 18–40.

Pascual-Leone, J. (1987). Organismic processes for neo-Piagetian theories: A dialectical causal account of cognitive development. *International Journal of Psychology, 22*, 531–570.

Pauli, W. (1925). Über den zusammenhang des abschluβes der elektronengruppen im atom mit der komplexstruktur der spektren. *Zeitschrift für Physik, 31*, 765–785.

Pauling, L. (1931). The nature of the chemical bond. Application of results obtained from the quantum mechanics and from a theory of paramagnetic susceptibility to the structure of molecules. *Journal of the American Chemical Society, 53*, 1367–1400.

Pauling, L. (1939). *The nature of the chemical bond and the structure of molecules and crystals*. New York: Cornell University Press.

Pauling, L. (1952). Quantum mechanics of valence. *Nature, 170*, 384–385.

Pauling, L. (1964). *General chemistry* (3rd ed.). San Francisco: Freeman.

Pauling, L. (1980). Prospects and retrospects in chemical education. *Journal of Chemical Education, 57*, 38–40.

Pauling, L. (1992). The nature of the chemical bond – 1992. *Journal of Chemical Education, 69*, 519–521.

Pauling, L., & Wilson, E. B. (1935). *Introduction to quantum mechanics with applications to chemistry*. New York: McGraw-Hill.

Perkins, D. N. (2006). Constructivism and troublesome knowledge. In J. H. F. Meyer & R. Lands (Eds.), *Overcoming barriers to student understanding: Threshold concepts and troublesome knowledge* (pp. 33–47). London: Routledge.

Perl, M. L. (2004). The discovery of the Tau Lepton and the changes in elementary-particle physics in forty years. *Physics in Perspective, 6*, 401–427.

Perl, M. L., & Lee, E. R. (1997). The search for elementary particles with fractional electric charge and the philosophy of speculative experiments. *American Journal of Physics, 65*, 698–706.

Perrin, C. E. (1988). The chemical revolution. In A. Donovan, L. Laudan, & R. Laudan (Eds.), *Scrutinizing science: Empirical studies of scientific change* (pp. 105–124). Dordrecht: Kluwer.

Phillips, D. C. (1995). The good, the bad, and the ugly: The many faces of constructivism. *Educational Researcher, 24*, 5–12.

Phillips, D. C. (2005). The contested nature of empirical educational research (and why philosophy of education offers little help). *Journal of Philosophy of Education, 39*, 577–597.

Phillips, D. C. (2014). Email to author, dated Feb. 6, 2014, reproduced with permission.

Piaget, J. (1985). *The equilibration of cognitive structures: The central problem of intellectual development*. Chicago: University of Chicago Press.

Piaget, J., & Garcia, R. (1989). *Psychogenesis and the history of science*. New York: Columbia University Press.

Pickering, A. (Ed.). (1992). *Science as practice and culture*. Chicago: University of Chicago Press.

Pienta, N. (2012). What we do and don't know about teaching and learning science: The National Research Council weighs in on discipline-based education research. *Journal of Chemical Education, 89*, 963–964.

Polanyi, M. (1964). *Personal knowledge: Towards a post-critical philosophy*. Chicago: University of Chicago Press. first published 1958.

Porter, T. M. (1981). A statistical survey of gases: Maxwell's social physics. *Historical Studies in Physical Sciences, 12*, 77–116.

Raman, V. V., & Forman, P. (1969). Why was it Schrödinger who developed de Broglie's ideas? *Historical Studies in the Physical Sciences, 1*, 291–314.

Ramberg, P. J. (2000). The death of vitalism and the birth of organic chemistry: Wöhler's urea synthesis and the disciplinary identity of organic chemistry. *Ambix, 47*, 170–195.

Ramberg, P. J. (2015). That Friedrich Wöhler's synthesis of urea in 1828 destroyed vitalism and gave rise to organic chemistry. In R. L. Numbers & K. Kampourakis (Eds.), *Newton's apple and other myths about science*. Cambridge, MA: Harvard University Press.

Reese, H. (1982). A comment on the meaning of 'dialectics'. *Human Development, 25*, 423–429.

Reinmuth, O. (1932). Editor's outlook. *Journal of Chemical Education, 9*, 1139–1140.

Riegel, K. F. (1979). *Foundations of dialectical psychology*. New York: Academic Press.

Robinson, J. T. (1969). Philosophy of science: Implications for teacher education. *Journal of Research in Science Teaching, 6*, 99–104.

Rocke, A. J. (1978). *Chemical atomism in the nineteenth century: From Dalton to Cannizaro*. Columbus: Ohio State University Press.

Rocke, A. J. (2013a). Email to author dated October 30, 2013, reproduced with permission.

Rocke, A. J. (2013b). Email to author dated November 3, 2013, reproduced with permission.

Rocke, A. J. (2013c). What did 'theory' mean to nineteenth-century chemists? *Foundations of Chemistry, 15*, 145–156.

Rocke, A. J., & Kopp, H. (2012). *From the molecular world, a nineteenth-century fantasy*. Heidelberg: Springer.

Rodebush, W. H. (1928). The electron theory of valence. *Chemical Review, 5*, 509–531.

Rodgers, G. E. (1995). *Introduction to coordination, solid state, and descriptive inorganic chemistry* (Spanish ed.). New York: McGraw-Hill.

Rodríguez, M. A., & Niaz, M. (2004). A reconstruction of structure of the atom and its implications for general physics textbooks. *Journal of Science Education and Technology, 13*, 409–424.

Ruse, M. (2014). Book review of P. Hoyningen-Huene, Systematicity: The nature of science. *Philosophy of Science, 81*(2), 284–288.

Rutherford, E. (1911). The scattering of alpha and beta particles by matter and the structure of the atom. *Philosophical Magazine, 21*, 669–688.

Rutherford, E. (1915). The constitution of matter and the evolution of the elements. In *Address to the annual meeting of the National Academy of Sciences* (pp. 167–202). Washington, DC: Smithsonian Institution.

Ryan, A. G., & Aikenhead, G. S. (1992). Students' preconceptions about the epistemology of science. *Science Education, 76*, 559–580.

Sakkopoulos, S. A., & Vitoratos, E. G. (1996). Empirical foundations of atomism in ancient Greek philosophy. *Science & Education, 5*, 293–303.

Sawrey, B. (1990). Concept learning versus problem solving: Revisited. *Journal of Chemical Education, 67*, 253–254.

Sawyer, R. K. (Ed.). (2006). *The Cambridge handbook of the learning sciences*. New York: Cambridge University Press.

Schmidt, H. J. (1997). An alternative path to stoichiometry problem solving. *Research in Science Education, 27*, 237–249.

Schwab, J. J. (1962). *The teaching of science as enquiry*. Cambridge, MA: Harvard University Press.

Schwab, J. J. (1974). The concept of the structure of a discipline. In E. W. Eisner & E. Vallance (Eds.), *Conflicting conceptions of curriculum* (pp. 162–175). Berkeley: McCutchan Publishing Corp.

Segal, B. G. (1989). *Chemistry: Experiment and theory* (2nd ed.). New York: Wiley.

Shaik, S., & Hiberty, P. C. (2008). *A chemist's guide to valence bond theory*. New York: Wiley-Interscience.

Shapere, D. (1977). Scientific theories and their domains. In F. Suppe (Ed.), *The structure of scientific theories* (2nd ed., pp. 518–565). Chicago: University of Illinois Press.

Sheldon, S. (2013). Bohmian mechanics. In E. N. Zalta (ed.), *The Stanford encyclopedia of philosophy* (Spring 2013 edition). http://plato.standford.edu/archives/spr2013/entries/qm-bohm (consulted 3 May 2015).

Shulman, L. S. (1986). Those who understand: Knowledge growth in teaching. *Educational Researcher, 15*, 4–14.

Siegel, H. (2013). Review of Stathis Psillos and Martin Curd (eds): *The Routledge companion to philosophy of science* (pp. 729–731), 2008. New York: Routledge. *Science & Education*, 22(3), 729–731

Siegel, H. (2014). Email to author, dated Feb. 5, 2014, reproduced with permission.

Smith, E. F. (1925). Observations on teaching the history of chemistry. *Journal of Chemical Education, 2*, 553–555.

Smith, M. U., & Scharmann, L. C. (1999). Defining versus describing the nature of science: A pragmatic analysis for classroom teachers and science educators. *Science Education, 83*(4), 493–509.

Smith, M. U., & Scharmann, L. C. (2008). A multi-year program developing an explicit reflective pedagogy for teaching pre-service teachers the nature of science by ostention. *Science & Education, 17*, 219–248.

Smith, M. U., Lederman, N. G., Bell, R. L., McComas, W. F., & Clough, M. P. (1997). How great is the disagreement about the nature of science? A response to Alters. *Journal of Research in Science Teaching, 34*, 1101–1103.

Sommerfeld, A. (1915). *Munchener Berichte* (pp. 425–458).

Sommerfeld, A. (1916). *Annalen der Physik, 51*, 1–94, 125–167.

Spencer, J. N., Bodner, G. M., & Rickard, L. H. (1999). *Chemistry: Structure and dynamics*. New York: Wiley.

St Clair-Thompson, H. L., Overton, T., & Bugler, M. (2012). Mental capacity and working memory in chemistry: Algorithmic versus open-ended problem solving. *Chemistry Education Research and Practice, 13*, 484–489.

Stamovlasis, D., & Tsaparlis, G. (2012). Applying catastrophe theory to an information processing model of problem solving in science education. *Science Education, 96*, 392–410

Staver, J. R., & Lumpe, A. T. (1995). Two investigations of students' understanding of the mole concept and its use in problem solving. *Journal of Research in Science Teaching, 32*, 177–193.

Stinner, A. (1992). Science textbooks and science teaching: From logic to evidence. *Science Education, 76*(1), 1–16.

Stuewer, R. H. (1975). *The Compton effect: Turning point in physics*. New York: Science History Publications.

Styer, D. F. (2000). *The strange world of quantum mechanics*. Cambridge: Cambridge University Press.

Talanquer, V. (2013). Chemistry education: Ten facets to shape us. *Journal of Chemical Education, 90*(7), 832–838.

Tarsitani, C., & Vicentini, M. (1996). Scientific mental representations of thermodynamics. *Science & Education, 5*(1), 51–68.

Taylor, P. C. (2015). Constructivism. In R. Gunstone (Ed.), *Encyclopedia of science education* (pp. 218–223). Heidelberg: Springer.

Teo, T. W., Goh, M. T., & Yeo, L .W. (2014). Chemistry education research trends: 2004–1013. Chemistry Education Research and Practice (in press).

Thomson, T. (1825). *An attempt to establish the first principles of chemistry by experiments* (2 Vols). London: Colburn & Bentley.

Thomson, J. J. (1897). Cathode rays. *Philosophical Magazine, 44*, 293–316.

Thomson, J. J. (1907). *The corpuscular theory of matter*. London: Constable.

Thomson, J. J. (1914). The forces between atoms and chemical affinity. *Philosophical Magazine, 27*, 757–789.

Thomson, J. J. (Chair). (1919). Joint eclipse meeting of the Royal Society and the Royal Astronomical Society. *Observatory, 42*, 389–398.

Thomson, G. P., & Reid, A. (1928). *Proceedings of the Royal Society of London A, 117*, 601–609.

Tobias, S. (2009). An eclectic appraisal of the success or failure of constructivist instruction. In S. Tobias & T. M. Duffy (Eds.), *Constructivist instruction: Success or failure?* (pp. 335–350). New York: Routledge.

Tocci, S., & Viehland, C. (1996). *Holt chemistry: Visualizing matter*. New York: Holt, Rinehart and Winston.

Tolvanen, S., Jansson, J., Vesterinen, V.-M., & Aksela, M. (2014). How to use historical approach to teach nature of science in chemistry education? *Science & Education, 23* (in press).

Toon, E. R., Ellis, G. L., & Brodkin, J. (1968). *Foundations of chemistry*. New York: Holt, Rinehart and Winston.

Toulmin, S. (1958). *The uses of argument*. Cambridge, UK: Cambridge University Press.

Towns, M. H. (2013). New guidelines for chemistry education research manuscripts and future direction of the field. *Journal of Chemical Education, 90*, 1107–1108.

Towns, M. H., & Kraft, A. (2011). Review and synthesis of research in chemical education from 2000–2010. Commissioned paper for the NAS Board on Science Education Committee on status, contributions, and future directions of discipline– based education research. http://sites.nationalacademies.org/DBASSE/BOSE/DBASSE_080124#.UYJ-kpX07jI. Accessed July 2014.

Tro, N. (2008). *Chemistry: A molecular approach*. Upper Saddle River: Prentice Hall (Pearson).

Tsai, C.-C. (2000). Enhancing science instruction: The use of 'conflict maps'. *International Journal of Science Education, 22*(3), 285–302.

Tsai, C.-C., & Wen, M. L. (2005). Research and trends in science education from 1998 to 2002: A content analysis of publication in selected journals. *International Journal of Science Education, 27*, 3–14.

Tsaparlis, G. (1998). Dimensional analysis and predictive models in problem solving. *International Journal of Science Education, 20*, 335–350.

Tsaparlis, G. (2001). Towards a meaningful introduction to the Schrödinger equation through historical and heuristic approaches. *Chemistry Education: Research and Practice in Europe, 2*, 203–213.
Tsaparlis, G. (2014a). Linking the macro with the micro levels of chemistry: Demonstrations and experiments that can contribute to active/meaningful/conceptual learning. In I. Devetak & S. A. Glažar (Eds.), *Learning with understanding in the chemistry classroom* (pp. 41–61). Dordrecht: Springer.
Tsaparlis, G. (2014b). Cognitive demand. In R. Gunstone (Ed.), *Encyclopedia of science education* (pp. 1–4). Dordrecht: Springer.
Tsaparlis, G., & Papaphotis, G. (2002). Quantum-chemical concepts: Are they suitable for secondary students? *Chemistry Education Research and Practice, 3*(2), 129–144.
Tsaparlis, G., & Papahotis, G. (2009). High school students' conceptual difficulties and attempts at conceptual change. International Journal of *Science Education, 31*, 895–930.
Tsaparlis, G., & Zoller, U. (2003). Evaluation of higher vs. lower-order cognitive skills-type examinations in chemistry: Implications for university in-class assessment and examinations. *University Chemistry Education, 7*, 50–57.
Tsaparlis, G., Kousathana, M., & Niaz, M. (1998). Molecular-equilibrium problems: Manipulation of logical structure and of M-demand, and their effect on student performance. *Science Education, 82*, 437–454.
Ünal, S., Çalık, M., Ayas, A., & Coll, R. K. (2006). A review of chemical bonding studies: Needs, aims, methods of exploring students' conceptions, general knowledge claims, and students' alternative conceptions. *Research in Science & Technological Education, 24*(2), 141–172.
Ünal, S., Coştu, B., & Ayas, A. (2010). Secondary school students' misconceptions of covalent bonding. *Journal of Turkish Science Education, 7*(2), 3–29.
Van de Waals, J. D. (1873). *Over de Continuiteit Van den Gas en Vloeistoftoestand*. Leyden.
Van Dijk, E. M. (2011). Portraying real science in science communication. *Science Education, 95*, 1086–1100.
Van Dijk, E. M. (2013). Review of Paul Hoyningen-Huene: Systematicity: The nature of science. *Science & Education, 22*, 2369–2373.
Van Fraassen, B. C. (1980). *The scientific image*. Oxford, UK: Clarendon.
Van Spronsen, J. (1969). *The periodic system of chemical elements. A history of the first hundred years*. Amsterdam: Elsevier.
Vargas Llosa, M. (2010). Nobel prize in literature acceptance speech. http://www.nobel-prize.org/nobel_prizes/literature/laureates/2010/vargas_llosa-lecture, downloaded 11 Dec 2010.
Vesterinen, V.-M., & Aksela, M. (2013). Design of chemistry teacher education course on nature of science. *Science & Education, 22*(9), 2193–2225.
Vesterinen, V.-M., Aksela, M., & Lavonen, J. (2013). Quantitative analysis of representations of nature of science in Nordic upper secondary school textbooks using framework of analysis based on philosophy of chemistry. *Science & Education, 22*(7), 1839–1855.
Viana, H. E. B., & Porto, P. A. (2010). The development of Dalton's atomic theory as a case study in the history of science: Reflections for educators in chemistry. *Science & Education, 19*, 75–90.
Vickers, P. (2013). *Understanding inconsistent science*. New York: Oxford University Press.
Wan, Z. H., Wong, S. L., & Zhan, Y. (2013). When nature of science meets Marxism: Aspects of nature of science taught by Chinese science teacher educators to prospective science teachers. *Science & Education, 22*(5), 1115–1140.
Wartofsky, M. W. (1968). *Conceptual foundations of scientific thought: An introduction to the philosophy of science*. New York: Macmillan.
Weinberg, S. (1980). Conceptual foundations of the unified theory of weak and electromagnetic interactions. *Reviews of Modern Physics, 52*, 515–523.
Weinberg, S. (2001). Physics and history. In J. A. Labinger & H. M. Collins (Eds.), *The one culture: A conversation about science*. Chicago: University of Chicago Press.
Weingart, P. (2004). Between science and values. In P. Machamer & G. Wolters (Eds.), *Science, values and objectivity* (pp. 112–126). Pittsburgh: University of Pittsburgh Press.

Weisberg, M. (2007). Who is a modeler? *British Journal for the Philosophy of Science, 58*, 207–233.

Weisberg, M., Needham, P., & Hendry, R. (2011). Philosophy of chemistry. In E. N. Zalta (Ed.), *The Stanford encyclopedia of philosophy*. http://plato.stanford.edu/archives/win2011/entries/chemistry. Consulted on 1 May 2015.

Wheaton, B. R. (1983). *The tiger and the shark: Empirical roots of wave-particle dualism*. Cambridge: Cambridge University Press.

Wheland, G. W. (1952). Book review of valence by C.A: Coulson. *Journal of the American Chemical Society, 74*, 5810.

Wiechert, E. (1897). Ergebniss einer messung der geschwindigkeit der cathodenstrahlen. *Schriften der Physikalischokonomisch Gesellschaft zu Konigsberg, 38*, 3.

Williamson, V. M. (2014). Teaching chemistry conceptually. In I. Devetak & S. A. Glažar (Eds.), *Learning with understanding in the chemistry classroom* (pp. 193–208). Dordrecht: Springer.

Wilson, D. (1983). *Rutherford: Simple genius*. Cambridge, MA: MIT Press.

Windschitl, M. (2004). Folk theories of "inquiry:" How preservice teachers reproduce the discourse and practices of an atheoretical scientific method. *Journal of Research in Science Teaching, 41*, 481–512.

Wolpert, L. (1993). *The unnatural nature of science*. Cambridge, MA: Cambridge University Press.

Woodward, R. B., & Hoffmann, R. (1965). Stereochemistry of electrocyclic reactions. *Journal of the American Chemical Society, 87*, 395–397.

Worrall, J. (2010). Theory-change in science. In S. Psillos & M. Curd (Eds.), *The routledge companion to philosophy of science* (pp. 281–291). New York: Routledge.

Wu, H.-K., & Shah, P. (2004). Exploring visuospatial thinking in chemistry learning. *Science Education, 88*, 465–492.

Yuan, K., Steedle, J., Shavelson, R., Alonzo, A., & Oppezzo, M. (2006). Working memory, fluid intelligence and science learning. *Educational Research Review, 1*, 83–98.

Zeidler, D. N., Walker, K. A., & Ackett, W. A. (2002). Tangled up in views: Beliefs in the nature of science and responses to socioscientific dilemmas. *Science Education, 86*, 343–367.

Ziman, J. (1978). *Reliable knowledge. An exploration of the grounds for belief in science*. Cambridge: Cambridge University Press.

Zoller, U., & Tsaparlis, G. (1997). Higher and lower-order cognitive skills: The case of chemistry. *Research in Science Education, 27*, 117–130.

Zoller, U., Dori, Y. J., & Lubezky, A. (2002). Algorithmic, LOCS and HOCS (chemistry) exam questions: Performance and attitudes of college students. *International Journal of Science Education, 24*(2), 185–203.

Zumdahl, S. S. (1993). *Chemistry* (3rd ed.). Lexington: Heath.

Index

M. Niaz, *Chemistry Education and Contributions from History and Philosophy of Science*, Science: Philosophy, History and Education,
DOI 10.1007/978-3-319-26248-2

MIX
Papier aus verantwortungsvollen Quellen
Paper from responsible sources
FSC® C105338

If you have any concerns about our products,
you can contact us on
ProductSafety@springernature.com

In case Publisher is established outside the EU,
the EU authorized representative is:
Springer Nature Customer Service Center GmbH
Europaplatz 3, 69115 Heidelberg, Germany

Printed by Libri Plureos GmbH
in Hamburg, Germany